LIVESTOCK GUARDIAN DOGS

LIVESTOCK GUARDIAN DOGS

An Illustrated Celebration

Cat Urbigkit

Skyhorse Publishing

Skyhorse Publishing books may be purchased in bulk at special discounts for sales promotion, corporate gifts, fund-raising, or educational purposes. Special editions can also be created to specifications. For details, contact the Special Sales Department, Skyhorse Publishing, 307 West 36th Street, 11th Floor, New York, NY 10018 or info@skyhorsepublishing.com.

Skyhorse® and Skyhorse Publishing® are registered trademarks of Skyhorse Publishing, Inc.®, a Delaware corporation.

Visit our website at www.skyhorsepublishing.com.

10 9 8 7 6 5 4 3 2 1

Library of Congress Cataloging-in-Publication Data

Names: Urbigkit, Cat, author.
Title: Livestock guardian dogs : an illustrated celebration / Cat Urbigkit.
Description: New York, NY : Skyhorse Publishing, 2016.
Identifiers: LCCN 2016033457 | ISBN 9781510709102 (hardcover : alk. paper)
Subjects: LCSH: Livestock protection dogs.
Classification: LCC SF428.6 .U737 2016 | DDC 636.737--dc23
LC record available at https://lccn.loc.gov/2016033457

Cover design by Tom Lau
Cover photo credit: Cat Urbigkit

Print ISBN: 978-1-5107-7492-6
Ebook ISBN: 978-1-5107-0911-9

Printed in China

Contents

Preface

A celebration of working livestock guardian dogs is long overdue. This book exalts the dogs that live and work across wild landscapes, and have done so for thousands of years. Within these pages you won't find photos of well-groomed guardians with famous names aligned with champion lineages. Instead, you'll meet typical working guardians from around the globe—dogs that face fierce predators so that their flocks are not met with harm.

Livestock guardian dogs are often promoted as a nonlethal means of deterring predators, which is only partly true. Most encounters between guardian dogs and predators result in the dog's disruption of the wild predator's hunting behavior, with the predator fleeing the encounter. That is the case with *most* encounters, but not all. In other cases, livestock guardian dogs kill predators—be they coyotes, bears, or wolves. At times, it's a matter of kill or be killed.

Guardian dogs are canine warriors to those that would dare threaten the guarded. Yet it is the dog's deep devotion to a weaker species that is the motivation driving the use of such physical force. To watch a guardian as it gently noses a newborn lamb is to witness a perfect link in nature's evolutionary chain. Several centuries ago, a naturalist impeccably described the guardian dog as "the only animal born perfectly trained for the service of others."

While this volume barely begins to address the complexities and accomplishments of the working livestock guardian, it is my humble offering in honor of this noble beast.

CHAPTER 1

NEVER PERFECT

The author's domestic sheep flock grazes in private pastures in the southern Wind River Mountains in the fall of 2013.

The wolves arrived in the darkest hour of night. They traveled through the rain, tongues lolling from open mouths, trotting swiftly on long legs, toes splayed in the mud as they gripped for traction, leaving massive paw prints behind as the only sign of their presence. Easily clearing the pasture fence in one powerful leap, the smaller gray female and the large black male were intent on returning to the sheep flock for an easy meal.

The wolves had hit the herd the night before, again under the cover of darkness. After spending the day dealing with the carnage of seven dead sheep, and two others that were so severely wounded only a bullet to the head could end their suffering, I had prepared for the wolves' return. Heavy overcast skies had left the valley shrouded in a gray gloom while I worked with Hud, one of our herding dogs, to bunch the sheep herd against the far corner fence, hurrying as the last rays of light dipped behind the mountains. Hud and I sat side by side on the ground facing the herd in a soft drizzling rain, waiting for the sheep to settle as darkness fell.

Rena was there to meet them when the wolves leapt into the pasture. One-hundred thirty pounds of determined Akbash sheep guardian dog, she met the wolves head-on, brawling in the distance from the herd, in the darkness, in the rain. When the wolves attacked, Rena could face one, as the other attacked her rear. The wolves sunk their teeth into her haunches, nearly severing her tail at its base and biting her tender underside. Fighting for both her own life as well as the lives of her sheep, Rena battled on, keeping the wolves from reaching the herd. The dog warrior returned successful, staggering as she brought her wounded, bloody body back to collapse at my side just before the sun began rising over the jagged granite face of the Wind River Mountains.

It was September 2013, and I had been camped with the herd as it grazed a series of private pastures in the foothills of Wyoming's southern Wind River Mountains—

Rena at work on the sagebrush steppe of Wyoming, prior to her battle with wolves.

pastures that have been used by domestic sheep flocks for more than a hundred years. Our small herd was protected by the three burros that are always present, and by a group of livestock guardian dogs. The range here is fluid and complex, with thousands of sheep and their guardian dogs coming and going, as well as the shepherds that accompany them. The sheep flocks carry the same genetics, and I've raised many of the guardian dogs that use this rangeland that covers countless square miles, some of which is divided into pastures, while others are allotments that include public land.

That weekend there were about a thousand sheep grazing in a two-mile area, with at least six livestock guardian dogs. The sheep spread out to graze during the day, but bunch up together to bed at night.

Where each dog was located with any bunch of sheep at any given time was fluid. We'd had a lot of bear activity, and the dogs had done a fantastic job of keeping the bears out of the sheep in this area.

One night a few weeks prior, when a bear got into a nearby cattle herd, two of the dogs from my bunch raced to the rescue, as did another guardian dog that came from the south. The two dogs returned to my bunch within about forty-five minutes, and the third dog returned to its station to the south. Rena, our five-year-old Akbash female, had stayed with my sheep—the only reason I know this much is because I was sleeping on the ground next to the flock that night, as I often do. I had believed that if we had problems in the sheep, it would probably be with wolves. When both black bear

A young Akbash guardian perches on a high spot to watch over the flock.

tracks and grizzly bear tracks were found the next morning, I gave up sleeping under the stars, and started using a tent as a more visible sign of human presence. The only wolf tracks that were found were old, but it quickly became evident that bears were a constant presence. Two of the Akbash dogs with my flock were particularly enthusiastic at hazing bears away.

The day before the wolves attacked my sheep, there was a combination of Akbash and Central Asian Ovcharka dogs guarding the herds, which were divided into two pastures, with the sheep bedded close to each other, but with a fence line between two main bunches. Some of the dogs were back and forth on patrol, and others stayed inside the flocks. I don't know what happened that night, since I wasn't there and the herder camped on the hill to the south couldn't see anything in the dark. When my husband Jim and I arrived on the rangeland to check our sheep the next morning, we found dead sheep, and walking wounded sheep that later had to be put down. One dog was missing, but later returned. The big herd in the adjacent pasture had been moved into the next allotment (still abutting my bunch), and we made plans to move the next morning since there wasn't enough time that day to do all that needed to be done. We checked the herd and hiked around both inside and outside the pasture, trying to find all the dead and wounded sheep, looking for tracks, and covering up some of the carcasses lest they be destroyed by ravens or other scavengers. In total, there were nine dead sheep (two ninety-pound

Rena's mother, Luv's Girl.

lambs and seven adult ewes weighing about 175–200 pounds each). I notified federal wildlife officials that we had a problem and needed an on-the-ground assessment. That would happen at first light on the next day, Monday. Jim was due back to work on Monday, so we raced forty miles back to the house to drop him off, and I threw my gear in the truck, loaded Rena and Hud, and went back to the herd.

After bunching the herd in the corner that evening, I parked my truck about a hundred yards from the flock, and slept in the cab with Hud, with the window down so I could hear and check the sheep through the night. The last thing I saw that night with my spotlight was Rena patrolling from the truck toward the far end of the pasture where the kills had happened the night before. I knew the burros were on that end as well, but couldn't see that far with my spotlight. I could see Luv's Girl (Rena's nine-year-old mother) sleeping with the flock in front of me. It was very dark, with the drizzle from the rain and the clouds completely hiding the crescent moon that was finally visible a few hours before sunrise. There were a few ruckuses during the night, and I could hear guardian dogs barking in various directions at infrequent intervals. Only once during the night did I see the sheep stand up from their beds in alarm, but Luv's Girl was still visible in front of them. They settled down and I went back to sleep. About 4 a.m., I let Hud out of the cab, and found wounded Rena sleeping on her side next to the truck, with fresh blood on her rump. I talked to her and she responded, but quickly went back to sleep. I spotlighted the sleeping herd, with Luv's Girl still present, and waited for daylight.

When darkness started easing, I could hear the neighboring sheep herd as the ani-mals started rising from the hillside to the south and I saw two Ovcharka guardians between that flock and my bunch, as well as Luv's Girl still bedded with my herd, and Rena next to the truck. I went to meet an approaching pickup truck, and within minutes, a Wildlife Services airplane flew over, breaking through the morning fog, shooting two wolves as they fled to the east. From her wounds, we know that Rena had fought with wolves during the night, but the wolves never made it to our flock because of her efforts. We don't know what role, if any, the burros or the other dogs played.

When Rena tried to stand and walk, she labored to work her back legs, and I could see the deep punctures through her long hair and undercoat. I did not try to assess further damage, but backed the truck up to a nearby ditch bank so she could load up, with me helping to lift her back end, and Rena crying out in pain as I lifted her. She collapsed in the bed of the truck, into an exhausted, wounded sleep.

There was a flurry of activity from that point, much of it involving other people coming to the rescue while I turned my attention to Rena and getting her to the vet clinic an hour away—my flock was moved to the south, and the carcasses picked up while I was driving the sixty miles to Pinedale. There would be at least two other large range herds exiting the mountains in the next few days (each flock with up to eleven guardian dogs), and they had been slated to rest and transition in these private pastures for a few days before they would begin trailing to the south. The plans were changed to speed the herds through this area. I learned that another guardian dog was brought off the mountain with his throat mangled by a predator. He did not survive.

Moving a flock from the high country of the Wind River Mountains.

Finally within range to pick up a cell phone signal, I called the vet clinic, warning them I was headed their way with a dog that had been injured by wolves. Brent, our veterinarian, had known Rena all her life, and knew the life-threatening danger posed to the dog from wounds inflicted by wild predators. By the time I got to town, Rena was unable to get up on her own. I climbed into the back of the truck and lifted her onto the gurney with the help of two other women from the vet clinic. Rena was very sweet about it all, stoic as she was strapped down and rolled inside the clinic. The clinic staff let me stick around for a few minutes to comfort Rena, as they checked her vitals and began preparing a plan of action. They

would administer fluids and sedation to her first, before shaving her to find the damage underneath.

When I returned to see Rena in the late afternoon, she was still groggy from sedation, but her dozens of wounds had been cleaned and stapled. She wagged her tail at me, which gave me hope that she'd pull through. The vet clinic staff said that the matted hair on her hind end had probably saved her life. Much of the hair had been pulled out, but it cushioned the bites. A few days later the clinic released Rena to come home, where I could help keep her wounds open and draining, and feed her painkillers and antibiotics. We had a few tense days with her, unsure if she would survive. Rena

Livestock Guardian Dogs

Five-year old Rena recovering from her conflict with wolves.

did substantially recover, but with some lingering stiffness in her hind end. Within three months she had returned to aggressively pursuing predators near her herd, with somewhat renewed vigor about it. She was five years old, splendid in her elegance, strength, and beauty.

When large carnivores and livestock share the same range, some animals will be killed—some wild, some domestic. It happens across western rangelands routinely, and while it is not pleasant, it is reality. It's hard to imagine what the damage would have been to our herd had our livestock guardian dogs not been present. We don't live within the full-time range of a wolf pack, so encounters with these animals are random events. Our encounters with coyotes are daily, and black bear encounters are frequent, but predator events are never entirely predictable.

Livestock protection is not an automated system, with predictable events and outcomes. There is no magical number or breed of dogs, or combinations of breed, age or sex, or set acreage, or fence design, or terrain, that allows a livestock producer to follow a formula to safeguard a herd from predation. I don't know how many black bears, grizzly bears, and wolves are in each area where our flock grazes, let alone how many of these animals the guardian dogs come into contact with. This is a fluid system, with livestock, predators,

Never Perfect

A guardian dog blocks the roadway while the flock passes.

and guardians sharing rangeland at random times and spaces, sometimes in conflict, but with varied outcomes. It's never perfect, but it works really well most of the time. It works because the guardian dogs serve as the brave buffer between predator and prey.

CHAPTER 2

LIFE WITH TUCKER

A goat herder in Turkey with his livestock guardian dog.

The scene seems nearly as old as time itself: A solitary shepherd, wooden staff in hand, tending to a flock of wooly ewes, with a gentle burro, its withers marked by the cross, trailing along behind. As the shepherd strides, a pair of massive guardian dogs move steadily at the edge of the herd, serving as its border guards. This scene comes from at least six thousand years ago in Central Asia, and from our western Wyoming ranch today.

Pastoralists around the globe have used livestock protection dogs to guard their herds from predators for thousands of years. The names of the breeds or types may vary, as do the people who use them and the type of livestock guarded. Regardless, guardian dogs have proven their worth to pastoralists time and time again.

We were launched into the world of livestock guardian dogs several decades ago, after starting a small sheep flock on an old ranch we had leased in the sagebrush of western Wyoming. A friend and mentor explained that with the relatively high density of predators in our area, we needed a flock protector to keep our herd safe. Thus, our first livestock guardian dog came into our lives.

Tucker was a young adult male Great Pyrenees. He had grown up on a ranch several hundred miles away, and his owner

A shepherd with a young Karakachan guardian dog in Bulgaria.

was seeking to rehome him after the dog had begun roaming to seek out predators. We live in a sparsely populated area, with our nearest (and only) neighbor located on a cattle ranch several miles away, so we thought this might be a good match.

The evening we brought Tucker home, he lazily wagged his tail in greeting at the orphan lambs in a pen, and we tethered him nearby for the night. That was our first night of trying to sleep to the sounds of the

deep, booming barks of a livestock protection dog on duty. There is danger in the dark, and the dogs must issue commands for the danger to remain at bay. Those barks have become such a part of my life it would be hard to imagine trying to sleep *without* that sound now.

Since Tucker was already an old hand at guarding sheep, we had no qualms about introducing him to the lambs the next morning. The white giant entered the lamb pen

Gentle with both sheep and cattle, Tucker (an adult male Great Pyrenees) liked to range for long distances—a trait that eventually earned him the nickname of Tourista.

without hesitation, gently waving his feathery tail over his back, massive head reaching out to gently lick the lambs in introduction. This was the first guardian dog the lambs had encountered, and their immediate response was one of contentment, with no fear of their big new companion.

Great Pyrenees dogs originated as the "great dogs of the mountains" guarding the flocks of Basque shepherds in the rugged Pyrenees of southwestern Europe. The breed is known not only for its large size but accompanying calm and affectionate temperament, as well as devotion to their flocks. Tucker lived up to his breed heritage, soon spending all his time with the lambs, resting outside their pen at night, and slowly

walking with the small flock as they left the pen for grazing during the day.

Tucker quickly became the flock leader. When I opened the gate to the pen in the mornings, Tucker greeted his lambs before walking them out to the meadow or highlands for grazing. The lambs followed, with the dog always nearby, and returned to the sheep shed in the evenings under his escort. Tucker eventually began jumping the fence that secured the lambs at night, so his visits to the lambs were completely unhindered by obstacles. Tucker was a devoted guardian, sometimes sleeping with the lambs' bodies snuggled in close to his.

Even though Tucker was devoted to his sheep, if Jim or I walked into the pasture to spend time with the flock, Tucker seemed to take that as a signal that he was free to leave, and always headed off in the last direction he'd heard a coyote. He never left the flock unless we were present to relieve him.

We entered the winter that first year with a flock of fifteen six-month-old ewe lambs, and Tucker as their guardian. The relationship between sheep and dog had become so close that we never doubted the safety of our small flock, which was remarkable considering where we live in western Wyoming. Our back fence line is the southern border of the Mesa, a big game winter range where human presence is restricted, and which serves as a coyote refuge.

As winter set in, we began to feed hay to the lambs, but also let them out during the day to graze on native vegetation that covers the ranch. One snowy January day, Tucker led his flock out for the day's grazing as usual. I checked on the sheep several times during the day as they traveled within their one-square-mile pasture. Late in the afternoon, it began to snow, and the sky quickly darkened. I went out to retrieve the flock, but could not find them. I became frantic as the snow began falling harder, and darkness began to spread across the range. Jim and I used spotlights to search in the darkness, but the snow was falling too hard for us to locate the missing flock.

Crestfallen, I had to leave the next morning on a week-long business trip that couldn't be delayed. Jim assured me that he would take time off from work away from the ranch and find the flock. It had snowed all night, so tracking wasn't an option, and it was evident the sheep were no longer in the immediate vicinity. Jim worked through a process of elimination to determine which direction the sheep might have gone, but had few clues to lead him. They weren't anywhere visible from a road, and they hadn't gone into the river bottom. His best guess was they had gone onto the Mesa—where the predators live.

I called home each day and was astounded and heartbroken to learn that our flock hadn't yet been found. Wherever they were, at least the lambs could graze, but Tucker must be starving. Jim's efforts to find the herd were frustrated by the fact that he couldn't use a vehicle to search, since the Mesa was closed to vehicles in winter. He chartered a private airplane to fly the Mesa, but finding fifteen one-hundred-pound white lambs with one white dog in a rough, snow-covered landscape proved futile.

I returned from my trip to an unhappy home. Jim and I were racked with worry and guilt about the missing flock and its guardian. We would begin anew the next morning, tossing and turning through another long, cold night.

With the morning sunrise came a ringing telephone. Verla, a neighbor seven miles away—on the distant side of the Mesa—called to ask if we were missing some lambs and a dog: they had just wandered into her ranch yard, eating hay and acting as though they belonged. When Jim pulled in with the stock trailer, the lambs happily jumped in, as though they would enjoy having a ride. Verla admitted that although she didn't like dogs, Tucker might be an exception.

I wept with relief. It had been seven days. Tucker had not lost a single lamb and seemed no worse for wear. He was hungry, but not ravenous.

That adventure would begin our process of constant learning about livestock protection dogs. As I watched Tucker and his flock in the coming days, I saw him catch and eat jackrabbits, which were abundant on the sagebrush rangelands that winter. Tucker moved with remarkable speed for such a large dog, resolving the mystery of how he'd survived so well when he'd taken the herd on its gallivant across the Mesa.

One sunny afternoon the next spring, I looked out the laundry room window to see Tucker barking vigorously from the edge of the herd. As I watched, I saw a coyote stalking toward Tucker, yowling and barking. The two soon met in battle, with the coyote lunging forward to bite Tucker's throat. Tucker swung his shoulders around, swinging the coyote back and forth, but not biting the little devil. The coyote had a firm grasp on Tucker's thick neck mane, but was unable to do the dog any harm. I was jumping up and down watching from inside the house, screaming for Tucker to "Kill! Kill!" Tucker used one big foot to slap the coyote off his neck and roll him across the ground in front of him. It became a standoff. The coyote stood a dozen feet from the dog, yowling and complaining in rage, while Tucker sat on his haunches calmly facing the coyote, refusing to allow the predator to approach the flock, which was now grazing away from the conflict zone. Tucker barked a warning to the coyote every now and then, and eventually the coyote skulked away.

Frustrated at his lack of aggression, I walked out to Tucker, and he greeted me in his always-friendly way. I couldn't be mad at him, but I decided that I needed a more aggressive dog. Tucker's failure to bite or kill a single coyote might not be a problem in this incident, but if a single coyote could keep him distracted, its sneaky mate might succeed in killing sheep during such encounters, while the big dog was otherwise engaged.

Our friend Pete comes from a family with a hundred-year history of sheepherding in the United States, and many more years before that in "the old country," as this Basque sheepman refers to his ancestral land. Pete's sheep outfit is huge, with thousands of sheep that walk and graze a seasonal migration route more than 125 miles in length. To keep the flocks safe, Pete keeps about twenty adult livestock protection dogs working at any one time. The dogs in Pete's migratory sheep outfit vie for breeding rights, and provide a classic example of survival of the fittest. I told Pete of my dilemma of Tucker not killing coyotes, and he sent me a beautiful adult female Akbash dog to try out. Tucker took one look at this spayed female we named FloJo and was smitten.

Originating in Turkey, the Akbash is narrower and sleeker in the flank than the Great Pyrenees, suggesting perhaps there was some sight hound in their origin since these guardian animals are fleet of foot and built to travel. Insistent on running down and killing any coyotes that approached the herd, FloJo provided an immediate improvement in predator control. Tucker followed along for sport, and the two became inseparable.

The guardian pair developed their own techniques for hunting coyotes. They would take the sheep out to graze along the irrigation ditch, and then hide in the brush nearby, apparently knowing that coyotes frequently use the brush cover along the ditch to approach the herd unseen. The coyote would do as expected, popping out into the open near the flock, and the dogs would give chase. Jim and I believe the dogs used the sheep as bait—that this was premeditated action. We've witnessed similar behavior in other guardian dogs over the years.

Threats to our flock became few and far between. Our local coyotes either got smart or got killed, so the dogs had to range farther for the hunt. The guardian pair started spending more time actively hunting coyotes than guarding their flock. We watched as their hunting loops got wider and wider, as they worked to clear coyotes from the range. An energy industry worker called to tell me about the pair's whereabouts one day—they had gone ten miles across the sagebrush rangelands in just over an hour.

Meanwhile, Pete told me he had a problem. One of his female guardian dogs had given birth to pups in the wild, and the wild pups needed to be caught and tamed. Jim and I drove forty miles south, where Jim and a herder dug the two wild and scared pups out of a culvert. We bundled the two-month-old pups up in our coats and took them home. The male pup was very ill, so we shuttled him to our veterinarian an hour away, where he stayed while he fought for his life against parvovirus. Our vet coined a name for the pup that he would eventually live up to, calling him Tick (a.k.a., The Incredible Coyote Killer). I took the more traditional route, naming the female pup at home Juel, after a geographic feature near where she was born. Within a few days, a still-weakened Tick was able to return home to us, and we kept the two pups in the house to try to tame them. Their mother was a very shy and wild dog (as were many guardian dogs working the western rangeland in those days), and her pups were the same. It took a great deal of persistent and quiet effort, but the shy pups gradually began to trust us. We spent hours sitting in our living room with the pups "hiding" in the front of our sweatshirts—holding those scared pups against our bodies until they realized if they were with us, they were safe. It was the earned trust of these two wild pups that made me fall in love with guardian dogs.

Once the pups had calmed enough that we could reliably catch them, we transitioned them outside to socialize with the adult guardian dog pair. Tucker and FloJo enjoyed romping with the young pups, and began taking them out to the sheep flock. This was good behavior, so we allowed it. Having an adult dog teaching young dogs is desirable. But the adult dogs were actually up to no good.

Eventually, Tucker and FloJo stole the pups one day and took them onto the Mesa. Jim searched and searched, walking through the snow, following the tracks as they headed up a rugged draw. It was snowing that day, with low visibility. Jim trudged on, eventually coming upon a fascinating scene. FloJo and Tucker, the white pair, apparently had some self-awareness. Standing off the side of the trail but facing

Tick and Juel as young pups, exploring the outside world, including following the ranch cat on its walkabout.

Jim, the dogs were barely visible. Seemingly aware that the only dark marks on their bodies were on their heads, the dogs turned their dark noses and eyes away from Jim's approach, allowing the camouflage of their white bodies to conceal their presence in the snow. Of course we can't know with certainty, but Jim is confident it was a deliberate act. The adult pair had placed the two pups into an old coyote den in the draw.

Jim once again retrieved the pups from their wild den, and carried the heavy burden back to the house through the snow. By this time, the pups weighed nearly thirty pounds each, and it was a long trek to the house. We never did figure out what the adult dogs intended to do with the pups. Although they had taken the pups from the safety of the ranch, they had deposited them in a safe den. Were the dogs reverting to wild ways, behaving as wolves do, or was there something we were missing? We'll never know.

We eventually agreed to do another dog swap with Pete, giving the adult pair a bigger range to roam. Pete took Tucker and FloJo for use with his migratory sheep flocks, where they thrived. Tucker eventually became known as "Tourista" among the sheepherders as he migrated with the range flocks, traveling back and forth for miles, clearing his ever-moving territory of coyotes, and breeding female guardians as they came in heat. The herders looked upon him favorably, admiring his travels as he trotted for miles on patrol, stopping in to visit each sheep herd, sometimes staying only for a few hours, or a few days. Other livestock protection dogs staying with the herds accepted the visits from the massive

A female Anatolian tending to her pups on Wyoming rangeland.

Tourista, but he did acquire a few battle wounds from disputes with those who begrudged his prowess. He lived a long and adventurous life, as top dog of an immense range.

In exchange for the adult pair, we took on a shorthaired female Anatolian-type dog named Seta, with her five tiny pups. Seta could provide some adult guardianship while Tick and Juel, nearly big enough to be coyote-proof, had a little more time to mature. It seemed like a good plan, but of course life would throw us another curve.

Tick and Juel, growing pups that they were, bumbled around the herd during the day, but had typically short puppy attention spans. One day they simply vanished. We couldn't find their tracks in the snow and weren't sure what direction they'd gone. We searched in all directions, to no avail. Day after day: nothing. We placed ads, hung posters, notified the sheriff's office and all the local veterinarians. Weeks passed. If the pups had gone onto the Mesa, coyotes must have killed them. At just four months old, they were too young to make it on their own, in winter. I finally gave up the search.

Good news often crosses the rural countryside via telephone call. One evening when it was just starting to get dark, our neighbors five miles upriver called to report that our pups had just appeared on their

Shy by nature, Tick was a Great Pyrenees/Akbash cross, born wild on the range. Although many Great Pyres are all-white dogs, others of this breed have badger, tan, grey, or reddish-brown markings.

deck. We made a mad dash to their place, and were greeted with the sight of our pups, now lanky adolescents with oversized feet, twenty-one days after their trip began. They were thrilled to see us as well, long-lost friends that we were.

The adolescent pups settled into a routine of guarding the flock and playing with Seta's new puppies. Tick worked out of his shyness and grew to be a large and beautiful dog. We sent him back to Pete, but retained Juel, who remained more shy and aloof with humans. My hours of cuddling Juel under a blanket or in the front of my coat had paid off: I was the only human she truly trusted. Although she adored Jim, she didn't entirely trust him.

Juel was magnificent with the animals. Her gentle soul was evident in the way she treated the sheep, especially the lambs.

Since she was about the same size as the adult ewes, sometimes lambs would follow behind her, mistaking Juel for their mother. Juel was always patient, even when those tiny lamb noses tickled her belly as they sought their mother's milk bag. Juel was careful to move slowly and not to knock the babies down while they were underneath her.

Juel was the first guardian dog that I took to a ewe in labor, knowing that Juel would watch over the ewe while she was in this most vulnerable position. The guardian's presence seemed to comfort the ewes. Juel would remain nearby, a silent sentinel, while the birth took place, and then consume the placenta and birthing material. This became an important task as our sheep flock continued to grow. Birthing materials are a predator attractant, but the livestock

Juel in her favorite place—the orphan lamb pen.

protection dogs quickly eliminate the danger, keeping our pastures clean. I've heard a few horror stories of livestock guardian dogs that couldn't be trusted during lambing for fear that they would kill lambs, but we have never had that experience. Our dogs are always well bonded and familiar with their sheep, and they most certainly know the difference between a live lamb and a pile of afterbirth. We don't allow adolescent pups in the lambing pastures without supervision though, since they like to play and roughhouse with each other, rather than being the quiet and calm presence the ewes need during that time.

Always gentle to all the sheep, Juel seemed to really enjoy being with them. I purchased a bunch of "bum" (orphan) lambs from a large commercial sheep outfit when Juel was just a yearling, and she cried until I let her into the pen with the small babes. That became Juel's favorite place, reclining in the pen among the smallest lambs on our ranch. The lambs curled up against her warm body, secure in her protection.

Juel befriended Seta and her five puppies as soon as the pups were old enough to begin to wander out of their doghouse. There was no aggression between the two female dogs, and their maternal instinct was a strongly shared characteristic. Juel helped to care for Seta's pups, though they were unrelated. Seta would leave her pups in Juel's care while Seta went out with the sheep flock during the day, returning every

A devoted sheep guardian, Juel even helped tend to the pups of other female guardians. Juel's reddish-brown facial markings were in contrast to her brother's gray markings.

Anatolians are a type of dog found in Turkey but developed into a breed in the United States, with a national breed club formed in 1970.

few hours to nurse the pups. Between the bum lambs in the pen, and supervising five growing puppies, Juel became the master babysitter. It suited her character.

Pete and I realized that exchanging dogs back and forth worked well for both of us. Our family's initial small stationary sheep flock allowed for the pups to be raised and well bonded to the sheep from a young age, and by the time I gave the pups back to Pete, they were large enough to survive life on the migratory sheep trail. In exchange, our flock was always well guarded, and we were able to have puppies in our lives on a regular basis. It's something we have done for more than two decades.

A female Great Pyrenees is straddled by a playful crossbred pup. Great Pyrenees dogs often have colored markings on their ears and faces.

CHAPTER 3

DOG OF NATURE

Range guardian dogs reigning at the head of the flock.

" This animal, faithful to man, will always preserve a portion of his empire and a degree of superiority over other beings," wrote eighteenth-century naturalist Georges-Louis Leclerc, Comte de Buffon, in describing a sheep dog in his thirty-six-volume *Natural History*. Leclerc continued, "He reigns at the head of his flock, and makes himself better understood than the voice of the shepherd.

"Safety, order, and discipline are the fruits of his vigilance and activity. They [the sheep] are a people submitted to his management, whom he conducts and protects, and against whom he never applies force but for the preservation of good order." Leclerc's description is sometimes difficult to follow, but is incredibly insightful, as he continues:

If we consider that this animal, notwithstanding his ugliness and his wild and melancholy look, is superior in instinct to all others; that he has a decided character in which education has comparatively little share; that he is the only animal born perfectly trained for the service of others; that, guided by natural powers alone, he applies himself to the care of our flocks, a duty which he executes with singular assiduity, vigilance, and fidelity; that he conducts them with an admirable intelligence, which is a part and portion of himself; that his sagacity astonishes at the same time that it gives repose to his master, while it requires great time and trouble to instruct other dogs for the purposes to which they are destined—if we reflect on these facts, we shall be confirmed in the opinion that the shepherd's dog is the true dog of nature, the stock and model of the whole species.

Modern scientists continue to debate the origin of the domestic dog—when and where the split from wolves occurred, whether it did so at multiple locations instead of one, and what prompted the domestication process. The debate will continue, and while fascinating, it is of little consequence to those living and working alongside our canine partners today.

Although described as harboring an ugly, wild, and melancholy look, the guardian dog was deemed to have superior instincts.

Guided by natural powers, a livestock guardian dog on the western range.

Dog fanciers also find disagreement when it comes to defining and describing dog breeds, and livestock protection breeds are no exception. Dog breed registries and associations seek to preserve their views of the "standard" of the selected breed, and do so with mixed results. I don't mean to discount the good work of these associations, but there is also much debate about the selection and exaggeration of certain traits within a breed, and that closed registries and defined standards fail to recognize the diversity within a specific breed. Regardless, the dog industry is a major economic force, and show champions can demand top dollar for breeding rights and pup sales. There are positive and negative aspects of the dog breeding and showing worlds—from trying to ensure quality dogs that are

humanely reared for potential buyers, to the opposite extreme of creating animals that cannot procreate on their own, and selecting animals for traits involving appearance instead of function.

Dog breed definitions and standards are of little consequence in the world of working guardian dogs on the landscape. What we can say with certainty is that livestock guardian dogs have been used for thousands of years to protect stock from predators across the broad landscapes of Europe and Asia. There are dozens of described breeds, but I tend to view these breeds as parts of larger land races, well suited to local conditions with much diversity and exchange throughout, and where natural elements have played a more prominent role than has human influence.

A rangeland guardian dog watches over a flock of fine-wooled sheep. Guardian dogs are independent thinkers.

The livestock guardian dogs that have long helped herdsmen protect domestic livestock from predators are generally of large sizes and calm dispositions, and are independent in nature. While their large size is subject to much attention, their calmness and independence attract little consideration. Independence is perhaps the guardian dog's most important trait, and is a reflection of the more primitive nature of the dogs—a trait they share with the wolves from which they descend.

Some standardized dog breeds (whose breeding and selection are human-controlled) seem to want nothing more than to please their owners. These human-dependent dogs make for pleasant, highly trainable companions. But working live-stock guardian dogs do not fall within this category—their primary purpose is not to please their human handlers, and independence remains at their core. Working guardians are faced with numerous opportunities for independent problem-solving in their daily lives on the landscape, and their very survival is dependent on making the correct judgment. Each working guardian assesses an approaching threat to the flock, and must respond accordingly, most often in the absence of immediate human leadership. Working guardians are attentive and gentle to the herds they guard, while aggressively protecting them from outside intruders.

When guardian dogs were introduced to the rangelands of the western United States, they became part of their livestock flocks,

A guardian dog leads a small herd of sheep down a country road in the American West.

much like guardians in the Old World. John Murray's 1878 book, *A Handbook for Travellers in Southern Italy*, described the multi-species social group of shepherd, flock, guardian dogs, and mules, with the Italian word *morra*. Guardian dogs now associated with western rangelands are a vital component of the *morra*.

Guardian dogs in the American West migrate with their flocks for hundreds of miles, mixing and interacting with other herds belonging to other livestock tenders. They are traded, stolen, left behind, rescued, and bought and sold—just like they were for eons along the Silk Road. Purebred guardian animals mix and breed with those of other protection breeds, producing yet another dog, often with a hybrid vigor. This is a natural breeding system in which the male dog most able to defend his selected female will win breeding rights (it is not necessarily the biggest, strongest male, but

perhaps the cleverest). The females select which male or males are most desirable to breed with, and they must be able to mate and whelp on their own in this natural system. Artificial insemination and caesarean surgery are not considered, and the nearest veterinarian may be one hundred miles away. It's a primitive system, largely unchanged from that of shepherds across Central Asia more than six thousand years ago. What matters most is the dog's ability to survive and do its job as a guardian, with all other considerations secondary.

For the most part, working guardian dogs around the world are not licensed, micro-chipped, or tracked with collars. Those who raise the dogs are not given permits from the government to do so, and don't have licensed kennels or websites promoting the dogs and their lineages. Their owners provide most of their needed veterinary care. There are exceptions to these

Recommended Reading

*SOS Dog: **The Purebred Dog Hobby Re-Examined**,* by Johan & Edith Gallant, 2008. For an extensive examination of purebred dog breeding, see this book written by Johan and Edith Gallant of South Africa.

Livestock Guardians: Using Dogs, Donkeys, and Llamas to Protect Your Herd, by Janet Vorwald Dohner, 2007. This informative book provides a concise summary of various livestock guardian dog breeds and their primary characteristics.

Livestock Guarding Dogs: Their Current Use World Wide, by Robin Rigg, 2001. Rigg prepared this comprehensive paper for The Canid Specialist Group. It provides a good overview of the dogs in most areas of the world. Search the internet for a copy of this often-cited paper.

Livestock Protection Dogs: Selection, Care, and Training, by Orysia Dawydiak and David Sims, 2004. This book provides a good introduction to those contemplating their first livestock guardian dog, whether to serve as a working dog or a human companion.

---

generalities, of course, but working guardian dogs are a vital part of overall livestock husbandry rather than a business venture for most livestock producers.

Although Spanish settlers to the American Southwest had brought guardian dogs (Spanish Mastiffs) into the region in the 1500s along with their fine-wooled sheep, their use was not widespread and long lasting in this New World.

John Hare Powell, in the *Memoirs of the Pennsylvania Agricultural Society* in 1826, describes the dogs imported with Merino sheep: "Their ferocity when aroused by any intruder, their attachment to their own flock, and devotion to their master, would, in the uncultivated parks of America, make them an acquisition of infinite value, by affording a defense against wolves, which they readily kill, and vagrant cur dogs, by which flocks are often destroyed."

An 1845 *American Agriculturist* article by J. H. Lyman described the "Mexican Shepherd-Dog" of New Mexico, descendants of the Spanish Mastiffs imported earlier. He wrote: "I have often thought, when observing the sagacity of this animal, that if very many of the human race possessed one half of the powers of inductive reasoning which seems to be the gift of this animal, that it would be far better for themselves and their fellow creatures."

In a 1985 paper, Ray Coppinger and his co-authors advanced three reasons why Spanish Mastiffs did not continue to be used in the American Southwest, including hybridization with other dogs; killing by soldiers; and the lack of knowledge of guardian dog traditions by English settlers in the 1800s.[1]

The use of guardian dogs died out with the rise of Anglo-American agriculture, according to researcher Jay Lorenz, as outlined in his 1990 doctoral dissertation. "Guarding dogs did not accompany the westward expansion of Merino stock since they were not a husbandry practice of the Anglo flock owners," Lorenz wrote. "The eventual dominance of Spanish sheep (Merinos and their derivatives) in the western United States was promulgated by the westward expansion of Anglos rather than a northern expansion of Spanish culture from the Southwest."[2] With major predators long gone from the New England states, and English settlers there not part of a guardian dog/transhumance livestock culture, guardian dogs were not present in these herds as they moved west.

With extensive private, local, and federal predator control programs instituted in the early 1900s in the United States—including the widespread use of poisons—many predator populations were severely reduced. Although coyotes still ranged across the landscape, as did a few mountain lions and black bears, some predator species were nearly eradicated. Gray wolves and grizzly bears were threatened with extinction, and were not judged to be near as significant a threat to livestock production as in years past. The poisons were indiscriminate in their victims, and dogs that encountered the poisons were killed as well.

But the 1970s ban on the use of predator poisons on public lands, and the con-

Navajo Dogs

One long-standing use of guardian dogs in the United States is that practiced by the Navajo on the Navajo Reservation (now known as the Navajo Nation) of Arizona, New Mexico, and Utah. Hal Black and Jeffrey Green published a paper in the *Journal of Range Management* in 1985 describing the Navajo dog: "Mixed-breed dogs of the Navajo appear to exhibit all behavioral traits believed to be important in protecting flocks from predators."[3]

These mixed-breed dogs weighed an average of thirty-seven pounds. Black and Green reported: "By habit and cultural tradition, the Navajo simply places the dog in an environment where imprinting (bonding) to the flock is obligatory. It is largely the absence of interaction between rancher and dog that creates a dog attentive to sheep." The paper also noted that some dogs were shy and appeared fearful of human handling or approach.

The two-hundred-year-old tradition of the Navajo involves raising the pups in the corral where the family's flock is penned every night. Children are not allowed to play with the dogs, dogs that leave the corral or herd are punished, and dogs that harass or injure the sheep or goats are killed. Most male dogs are castrated to reduce their interest in females, and to keep them from urinating around the homestead, since the houses and corrals are located in close proximity. The dogs bark at, chase, and occasionally kill coyotes (the common predator of the area). The dogs are fed dog kibble and/or scraps in communal dishes.

current listing of gray wolves and grizzly bears as protected species pursuant to the Endangered Species Act, meant that livestock producers needed to look at alternative methods to keep their herds safe. The possibility of renewing the use of livestock guardian dogs in the United States was contemplated, but there were several lingering assumptions hindering their adoption. Some believed that the dogs could only work with small flocks under fenced conditions, and that close personal attention by the stockowner was necessary. It has taken decades to get beyond those assumptions and to understand the value of guardian dogs in a variety of conditions across the United States.

Regardless, federal officials and several universities began placing guardian dogs on farms and ranches across the country. Guardian dogs of several breeds were imported into the United States under organized programs in the 1970s, and use of these dogs as a nonlethal method of predator control has since expanded.

The Great Pyrenees is a fairly common guardian breed and is easy to acquire in the United States, so it's not surprising that Pyr-

A Great Pyrenees issues a warning bark to predators within hearing range.

enees dogs became somewhat widespread throughout American ranches that were willing to try guardians to protect their flocks. Many of the dogs worked well, while others did not. Some were obtained from commercial breeders and had no recent history of livestock protection in their lineages, and others had various health or behavioral problems. Livestock owners began using the dogs without the benefit of cultural or historical use of the dogs, so their learning was largely dependent on a great deal of trial and error.

Research on the effectiveness of livestock guardian dogs in the United States in the early days of the program attempted to measure the effectiveness and attentiveness of the various breeds in use. Most of the work was done in association with Hampshire College in Amherst, Massachusetts, and the US Sheep Experiment Station in Dubois, Idaho.

A 1983 paper on the attentiveness of guardian dogs written by Ray Coppinger, Jay Lorenz, John Glendinning, and Peter Pinardi in the *Journal of Range Management* reported on the attentiveness of the progeny of dogs imported to the United States from Italy (Maremma), Turkey (Anatolian Shepherd), and Yugoslavia (Shar Planinetz). The dogs were placed with livestock producers throughout the United States beginning in 1978, and the Maremma and Maremma/Shar crossbreds were rated the most attentive.[4]

A later paper by Coppinger and his co-authors assessed the work of ten years of placing guardian dogs on working livestock ranches and farms. More than a thousand dogs were placed over ten years, in thirty-seven states. Guardian dogs reduced predation by 60 to 70 percent or more. According to the paper: "In the United States, the only places where dogs were judged not effective were those where sheep scattered widely over a great area and never flocked, or where producers did not spend more than a minimal amount of time with the flock."[5]

Jeffrey S. Green and Roger A. Woodruff published the results of a survey of sheep and goat producers in the United States and Canada in 1988. They found that 71 percent of respondents rated their dogs as very effective; 21 percent as somewhat effective; and 8 percent reported their dogs were not effective at deterring predation on their herds. Eighty-two percent of respondents viewed their dogs as an economic asset. No breed was rated more highly than others, and the rate of success was no different between males or females. Pyrenees and Komondors (Hungarian breed with a corded coat) were the most common breeds represented. According to the survey, more Komondors bit people than did Pyrenees, Akbash, or Anatolians, and fewer Pyrenees injured livestock than did Komondors, Akbash, or Anatolians.[6]

In 1999, William F. Andelt surveyed producers who used only one breed of dog in their operations, including Akbash, Great Pyrenees, and Komondor. The study did not find a difference in the rates of effectiveness of these breeds. But producers who used more than one breed rated Akbash as more effective than Great Pyrenees and Komondors in deterring predation.[7]

In both the Green–Woodruff and Andelt studies, Akbash dogs were rated most effective when all predators (including domestic dogs, bears, coyotes, and mountain lions) were considered. Akbash were reported to be more aggressive toward predators, as well as more active, intelligent, and faster than Great Pyrenees, while the Pyrenees were less likely to be aggressive toward unfamiliar domestic dogs.

An adult female Akbash showing attentiveness and interest to her sheep.

Mathieu Mauriès of France lives with ten Great Pyrenees dogs protecting his sheep and goats in the Alps of Haute Provence, and reports no livestock losses to his flock in more than a decade, despite the presence of large carnivores.

Researchers Inger Hansen and Morten Bakken tested Great Pyrenees dogs for possible use in Norway in the 1990s. Norway's 2.5 million sheep graze in the mountains and forested areas in the summer months, and an estimated one hundred thousand of the sheep "disappear" each year, with brown bears, lynx, and wolverines their major predators. The dogs were found to be very effective at chasing bears from the herds, and losses to lynx were reduced in the presence of these guardians. But reindeer herds overlap with sheep herds in seasonal grazing, and the researchers also found that the majority of the dogs chased reindeer. To resolve this conflict, the researchers recommended that the dogs also be exposed and bonded to reindeer early in life, so they could protect reindeer herds that are also subject to large carnivore depredations.[8]

Finland is a region without a history of guardian dog use. But once introduced, producers found the dogs eliminated livestock losses to protected herds. Primary predator threats to Finland's cattle, sheep, poultry, horses, and goat herds are from bears, lynx, and wolves. Finnish farmers also reported a feeling of security within farmyards and settled areas where large predators were considered a threat to human safety. Other benefits of the presence of livestock guardian dogs include the termination of damage to pasture fences due to elk, and the termination of damage to horticultural crops by deer. Finland (along with the other Nordic countries) has a unique situation in which "every man" has a right to roam for recreation and exercise, so guardian dogs will encounter joggers, skiers, berry pickers, hunters, etc., yet conflicts were not reported. Other areas of Europe have varied levels of this "freedom to roam," allowing public access to uncultivated and forested areas (including private property), and that poses special challenges and considerations for livestock protection dog owners.

The college and federal government programs placing livestock protection dogs on farms and ranches in the United States were gradually replaced by producers finding their own sources for dogs, and private breeders selling guardians became more widespread. Whether producers continue to use guardian dogs often depends on their initial success with the dogs, and the availability of quality dogs. The use of poor-quality dogs discourages the use of guardians, and producers who have an initial negative experience with the dogs may not continue to use guardian dogs even if quality dogs are available.

The guardian dogs used in the early days of the program in the United States were effective against the most common predator they encountered—coyotes. But the recent recovery of large carnivore populations in expanding ranges across the nation is causing a resurgence of interest in the use of livestock guardian dogs as producers are faced with new predators in their neighborhoods.

Mixed Breeds

Researchers discovered goat herders in Patagonia using mixed-breed, medium-sized dogs to protect their flocks. According to a paper by Alejandro González and co-authors, "large-bodied, purebred dogs are not practical for goat herders in northern Patagonia because these herders cannot afford them." So the local herders, with herds averaging about 730 goats, make do with the dogs that are available, rearing the pups by allowing them to be raised with the herd from a young age, and even suckling on does.

The researchers expanded on the limited use of mixed-breed guardians in a six-year pilot project by placing mixed-breed pups with other herders. Even though about half the dogs failed to bond with the goats (and some of the dogs ended up either not staying with the herd, attacked the herd, or tried to herd the goats rather than guard them), the project is considered a success because the dogs that did succeed were effective in reducing predator losses as well as reducing retaliatory killing of native carnivores by the herders.[9]

CHAPTER 4

BUILDING THE BOND

Lining the natal den with wool is a recommended first step in the bonding process.

The pups learn to associate the smell and feel of wool with the comfort of the natal den.

Key factors to successful livestock guardian dogs include acquiring dogs from working lineages; bonding pups to the species to be protected at an early age; and managing the dogs in a working partnership with the shepherd.

Getting livestock guardians off to a good start is an integral component in their future success. When a new litter of pups is born on our ranch, we often bed the pups in wool retained from our most recent sheep shearing. In doing so, we ensure that long

A runt guardian pup finds companionship in a pen of lambs. Penning a young puppy with a small group of animals it will later protect can be an important part of the bonding process.

before their eyes are opened to the outside world, the pups associate the smell and feel of the wool with the warmth, comfort, and security of their natal den. The pups are thus immediately drawn to the live sheep when they venture out a few weeks later, since the sheep carry the same smell and association.

The prime socialization period for livestock guardian pups appears to be from eight to sixteen weeks, and it's during this time that the pups should be introduced to the animals they are expected to grow up to guard—be it sheep, goats, cattle, chickens, turkeys, or any other domesticated, social species. Gentle supervision of young animals while they are getting to know their flock is necessary, and allowing a pup ample opportunity to bond with this new species is important.

It is much easier to bond pups to a sheep flock that has already had an association with guardian dogs than to a naïve flock. Sheep already accustomed to living with guardians show a natural curiosity toward new pups, as is evident when new pups arrive on our ranch and the adult sheep come to investigate. When the pups walk underneath the ewes, or sniff the underbelly of the rams, the sheep do not panic or become upset. They don't stomp the pups, but demonstrate a remarkable amount of patience as pups chew on the big curls of a ram's horn, or investigate a ewe's udder.

Building the Bond

When the pups venture out of the den, they are able to encounter sheep and are attracted by the familiar smell of wool. The pups can seek food and shelter away from the flock, but access should always be enabled during this bonding stage.

It is said that a good livestock guardian dog should be attentive, trustworthy, and protective of its flock. The behavioral clues of dogs that possess these behaviors are readily visible, as the dog lounges amid or moves with its flock; the dog engages in gentle exploration through licking or sniffing the sheep, or the dog rolls onto its back in an exhibition of submission; and the dog barks and retreats to the sheep in reaction to a perceived threat.

Livestock producers use a variety of bonding methods to ensure a guardian's future success. It's ideal for us if pups are born in the spring when the sheep flock is giving birth to its lambs, so the pups and lambs can be raised together. Since sheep are social animals, the young lambs welcome the companionship of the pups, while adult ewes supervise the process. The resulting bond formed from such an association is strong.

Other times we'll place the runt of a guardian dog litter in a pen of orphan lambs so it will have sheep companions to cuddle with, without the food competition from larger pups in the litter. As the pup grows, so do the lambs, but we always supervise to make sure the playful pup doesn't harass the lambs, or else we'll stick a few older sheep in the same pen to provide some discipline.

Guardian dogs that will be expected to protect multiple species should have access to those species at an early age. A bonding

A guardian pup showing proper submissive behavior to members of his flock.

pen may contain sheep, goats, chickens, and a few cows and calves. It's a good practice to change out the livestock on occasion so the pups bond to the species rather than to individual animals. It is also important to make sure that the pups can get away from the livestock to eat. Livestock should not be allowed to compete with the pups for food, since this can cause aggression in the dogs.

Louise Liebenberg raises Šarplaninac livestock guardian dogs on her family's ranch in Alberta, Canada. This medium-sized breed originated in the Macedonia/Yugoslavia region and is also known as Shar Planinetz or "Shars." Louise prefers that pups be born in the sheep barn, allow-

Some producers prefer to raise pups with non-playful stock such as calm rams. Close contact with livestock at this early age is when socialization and bonding occurs.

A pup investigates a gentle ram.

ing the pups to imprint to the sheep smells associated with that location. Once the pups start venturing out, they can interact with the sheep, but Louise never puts pups in a pen with orphan lambs, instead preferring to raise the pups "with non-playful stock such as quieter ewes or even friendly breeding rams." Once the pups are a few months old, they are transferred into a larger "boy's pen" where non-breeding rams, bulls, and horses are held, allowing the pups to become familiar with the ranch's varied livestock species.

Many producers use bonding pens to bond young dogs to the sheep, goats, or calves they are expected to guard. One Wyoming rancher lets the pups stay in the natal den with their mother until they are ready for weaning, and then places the litter of pups in a small pen with a group of gentle yearling rams, adult ewes, or ewes with lambs at their sides. The pups live in the small pen for a month or so, and are able to go underneath the hay and grain feeders for safety, but can venture around the livestock as they are comfortable. The larger livestock provide disciplinary action to the pups (usually by butting) when the pups try to roughhouse with the stock. The pups are fed and handled daily, becoming accustomed to human touch and attention. The pups are gradually transitioned out of the bonding pen. Once the pups weigh more than about thirty pounds (slightly larger than local coyotes), they can be released into a larger livestock herd, under super-

Adolescent pups in a bonding pen with young rams. These well-bonded, five-month-old pups are now getting their adult teeth, so they are ready to head out to the range to learn from and assist older guardians.

vision of older guardian dogs and their human herders. This ranch has several large range flocks that migrate hundreds of miles for grazing each year, and the strong bond formed in the bonding pen continues as the dogs join the large flocks on the migratory trail as adolescents. The pups are then exposed to other guardian dogs of various breeds, sexes, and ages, as well as herding dogs.

The bonding pen is located in a busy location near the ranch headquarters so the dogs are able to see and smell not just the sheep in the bonding pen, but the cattle and goat herds, saddle and draft horses, ranch machinery, and vehicles as work continues on the ranch around them. In doing so, the pups are exposed to a variety of stimuli that they will encounter later in life, reducing the possibility the dogs will experience fear when they encounter these objects at close range. Affection is not withheld from the pups, and most come when called, and receive meat treats for doing so. The rancher handles each dog intentionally, ensuring that when he needs to handle the dog for veterinary care (or for any other reason) the capture is not a stressful event for the dog. The dogs then perform well, in whatever situation they find themselves—at a remote range camp, ranch headquarters, shearing pens, or sorting corrals, etc. It's a system that works exceptionally well.

Building the Bond

A ewe chews her cud while a guardian pup naps.

There is no perfect model for how to raise effective livestock guardians. Just because one producer uses a certain system doesn't mean that system is superior—it's what works for that producer. While it may be tempting to pass judgment on those using different techniques, it's best to simply realize that individual producers can and should do what works best for them. Dogs are highly adaptable to a variety of conditions, as should be their human owners.

Organized programs to place guardians on ranches in the United States were accompanied by recommendations to producers, including an emphasis on the need to keep the dogs from becoming pets through a minimization of human contact. This later became a topic of great concern due to the management problems that followed. I believe that following this advice too closely led to generations of shy guardian dogs that stayed with their sheep but could not be caught by their owners. When these animals needed veterinary care, they suffered or had to be shot because of their owner's inability to capture and handle them.

Even today, USDA Wildlife Services continues to include such a recommendation in its educational materials. A 2010 fact sheet notes, "Rearing pups singly with sheep from a young age with minimal human contact is probably the most critical ingredient for success, allowing the dog to develop social bonds with the sheep."

The rearing pups "singly" may work well, but our experience does not indicate that it works better than bonding a group of pups with a group of sheep. We prefer allowing multiple pups to be raised together with a larger group of sheep, so when the dogs feel the need to play and blow off some steam, they can do it with siblings rather than with the sheep. We also prefer to raise pups in mixed-age groups as much as possible, so the pups can have the companionship, social interaction, and supervision of more experienced guardians.

The view of raising pups alone with their livestock is an old one, described in Charles Darwin's journals. Darwin recounted the shepherd dogs he observed in South America in his 1839 *The Voyage of the Beagle*, particularly noting the way the pups were reared:

While staying at this estancia, I was amused with what I saw and heard of the shepherd-dogs of the country. When riding, it is a common thing to meet a large flock of sheep guarded by one or two dogs, at the distance of some miles from any house or man. I often wondered how so firm a friendship had been established. The method of education consists in separating the puppy, while very young, from the bitch, and in accustoming it to its future companions. A ewe is held three or four times a day for the little thing to suck, and a nest of wool is made for it in the sheep-pen; at no time is it allowed to associate with other dogs, or with the children of the family. The puppy is, moreover, generally castrated; so that, when grown up, it can scarcely have any feelings in common with the rest of its kind. From this educa-

A guardian pup on a walkabout with a ram.

tion it has no wish to leave the flock, and just as another dog will defend its master, man, so will these the sheep. It is amusing to observe, when approaching a flock, how the dog immediately advances barking, and the sheep all close in his rear, as if round the oldest ram.

There is an argument to be made that individual livestock protection dog pups need varied levels of human attention as they grow and mature. Perhaps a bold and brave pup should not have a great deal of human attention, since the dog may later approach strangers for such attention. But a naturally shy pup will need much more human attention to bring it to the point that the shepherd will be able to catch and handle it as needed. Dogs that are somewhat in the middle of this range will need at least a little human socializing, so they are easily catchable but won't go wandering off with a friendly stranger.

Dog trainer Peggy Duezabou raises Akbash dogs on her Montana ranch, which is located in what is called the urban interface—an area where large ranches abut small privately owned parcels near town. Since she hosts herding dog lessons and trails, Peggy socializes and obedience trains her guardian dogs so she can have the best of both worlds—effective deterrents to predators, while not overly aggressive to dogs and visitors to the ranch. It's a process she's used very effectively.

As young dogs grow and become playful, adult ewes will discipline them if they become too rambunctious. The pups are butted until they lay quietly and show submission to the sheep. We've noticed that it works best if young dogs are paired with adult dogs, so they can learn appropriate behavior from the mature animals. Robin Rigg studied guardian dogs in Slovakia and noted another ben-

efit of having multiple dogs in a flock, reporting that two dogs in the same flock appeared to be "more confident, protective, and effective at confronting intruders than just one."[10]

J. H. Lyman's 1845 article in the *American Agriculturalist* also described how Spanish Mastiffs raised in New Mexico were bonded and raised:

The peculiar education of these dogs is one of the most important and interesting steps pursued by the shepherd. His method is to select from a multitude of pups a few of the healthiest and finest looking, and to put them to a sucking ewe, first depriving her of her own lamb. By force, as well as from a natural desire she has to be relieved of the contents of her udder, she soon learns to look upon the little interlopers with all the affection she would manifest for her own natural offspring. For the first few days the pups are kept in the hut, the ewe suckling them morning and evening only; but gradually, as she becomes accustomed to their sight, she is allowed to run in a small enclosure with them, until she becomes so perfectly familiar with their appearance as to take the entire charge of them. After this they are folded within the whole flock for a fortnight or so, they then run about during the day with the flock, which after a while becomes so accustomed to them, as to be able to distinguish them from other dogs—even from those of the same litter which have not been nursed among them.

Although the tradition of suckling a pup on a domestic ewe is indeed long-standing, it's not the only method used by producers around the world. We visited a farm in Bulgaria where litters of pups are raised and fed communally, surrounded by a variety

A litter of livestock guardian pups penned with ewes and their lambs. The presence of adult sheep as well as their lambs helps to ensure that the pups will be disciplined should they try to roughhouse with young livestock.

of tethered guardian dogs, and with other guardians ranging freely and moving back and forth with the flock as it was turned out to graze the mountains during the day, but returned for penning at night.

We also visited a cattle ranch in Spain where two three-month-old Spanish Mastiff pups were locked in a large barn with yearling cattle. The dogs were able to wander about the herd as desired, or could seek seclusion in several smaller pens not accessible to the cattle. The pups were handled and

fed daily by the ranch foreman, and played with by his children. Eventually the pups and cattle are released into larger paddocks outside, and continue to live together from that point forward, eventually joining with larger cattle herds in pastures on the ranch.

Another producer in Spain has large lambing barns where sheep are sorted into pens of varied size according to their status (about to lamb, already lambed, ewes with week-old lambs, etc.). Guardian dog bitches give birth in whelping boxes in the

Building the Bond

Akbash pups at rest with a ewe.

same building, and the pups are able to freely move about the sheep as they desire. As lambing season ends and the sheep flock goes to outside pastures, the dogs follow. The dogs are petted and fed daily by a variety of workers on the ranch.

We saw a variety of systems of raising pups in the rural areas of Turkey. In one case, the bitch digs a den into the side of a dirt bank near her grazing flock, and gives birth in this den. The pups start venturing out to meet the flock as they are able, and naturally grow up alongside, and amid, the herd they will eventually guard. The shepherd begins feeding and caring for the pups at an early age, once again ensuring that the animals are easily caught and handled later in life.

Other shepherds in Turkey create doghouses or whelping pens for their dogs, always in fairly busy locations amid their agricultural operations, so the pups are exposed to a variety of species and activities from an early age. While some of the dogs are not to be petted by visitors, the shepherd is always affectionate and able to handle his own guardian dogs.

Selecting a pup from a litter can be difficult. Some have a preference for a certain sex, while others select for size, presence or absence of back dewclaws, color, or other characteristics. Carla Cruz of Portugal notes

Livestock Guardian Dogs

A gentle adult ewe supervising puppy play. Experienced ewes will help to raise pups to behave properly in the flock.

that in her country, if a shepherd wants to pick the best pup of the litter, one method is to separate all the pups, using the "Mother Knows Best" rule: the pup the mother picks up first is the best.

Robert Vartanyan of Russia wrote that Caucasian Ovcharka pups are selected by shepherds based on the independent body language of a pup, including "a proud carriage, playfulness, and a desire to move

A livestock guardian dog raised on a ranch with both sheep and cattle is attentive to both.

about . . . A tail high set or even better curling over the back is a sign of a dog that is active and eager to work."

Some shepherds have color preferences. It is said that Italian shepherds prefer white dogs, in theory because it is easier to distinguish between dark-colored wolves and white dogs, so the shepherd won't kill the dogs in the heat of a fight. Shepherds in Portugal apparently prefer dogs with black muzzles, ears, and limbs, and with dewclaws, because dogs with these characteristics are the best guards and match the color of their sheep (which are white with black spots). Herders I know from Nepal select against red-colored dogs because dogs with that color are associated with viciousness in their homeland.

An unpleasant reality is that in their countries of origin, only one or two pups are typically raised from each litter. This heavy culling process helps to ensure that limited resources are not expended on numbers of dogs beyond the needs of the local shepherd.

Selecting males or females is simply a producer's preference, and there appears to be no significant difference in overall effectiveness. Of course, pregnant females and those raising young pups cannot be expected to be active guardians during these periods.

CHAPTER 5

GUARDIAN BEHAVIORS

A young guardian dog surges forward, barking in response to a perceived threat.

Livestock guardian dogs use a pattern of escalating behavior to keep predators from preying on their flocks. The first is territorial exclusion, in which the dogs actively patrol and scent-mark the area around the flock (whether it's stationary or migratory), claiming that territory from predators (especially other canines). Disruption is the second step, and involves the dog's barking and posturing in a position between the predator and the flock, serving to interrupt the hunting behavior of the predator.

Disruption is followed by confrontation, in which the dog approaches the predator, using intimidation and chase to force the predator's withdrawal. If the predator does not retreat, the confrontation can proceed until the dog attacks the predator, engaging in physical battle. Most predator conflicts with guardian dogs end at the confrontation stage as the predator retreats, but in areas such as our range, where predator populations are abundant, guardian dogs are frequently involved in physical battles with predators (especially coyotes, but sometimes also with wolves or bears). It appears that in areas with lesser predator problems, such incidents are more rare.

The dogs seem to pick their position by their individual natures, with one dog serving as the lead animal, forging out just ahead of the herd as it moves, while others stay near the flank of the herd, with another trailing along behind. Most of the dogs will join together to confront a perceived danger. Sometimes this means a group of large dogs barking and charging forward aggressively, but when the danger is perceived to be higher, the dogs will attack.

Guardian dogs that we've raised on our ranch exhibit varied types of protective behaviors. The first type was Tucker, a young adult Great Pyrenees male that worked alone by always staying with livestock, attempting to disrupt predators near the flock but initially not actively seeking out coyotes or other predators. He did not kill predators, but attempted to keep them out of the herd. Guardian dogs like this big, gentle dog make great farmstead dogs.

Two yearling females—Central Asian Ovcharka and Akbash—at play.

An adult male Central Asian Ovcharka hides from a vantage point above his herd.

Luv's Girl is an example of a second type. Actively occupying the canid niche, she displaces other canids within the same range, chasing and fighting predators invading her territory. This Akbash was athletic enough to be able to run down jackrabbits, and could displace and kill coyotes. She worked with another athletic young male Akbash to effectively harass both grizzly and black bears away from our flock. Perhaps most importantly, when multiple wolves were in the area near her flock, she did not actively patrol outside the herd, but remained close to the flock.

Rant, one of our Central Asian Ovcharkas, represented a third type involving both aggression and individualism. His high level of canine aggression made him a dangerous

threat to canine predators, and his enthusiasm for initiating attack—actively challenging predators—made him appropriate for guardian duty within wolf range. But this level of canine aggression could pose problems with other guardian dogs. His brother Turk was also an excellent guardian, and was far more amiable in terms of aggression toward other guardian dogs. Their sister Vega was an exceptional guardian, actively killing coyotes and displacing other predators, while more gregarious in terms of working with other dogs. Although she worked well in a dog pack protecting the herd, she did beat up other dogs until they submitted to her, acknowledging her dominance.

A fourth type involves dogs that both displace and actively hunt predators, seek-

Guardian Behaviors

A young male Akbash moves forward with raised tail to investigate a disturbance near the herd.

ing to kill them. Our female FloJo was this type, seeking out canids to kill.

Some dogs will combine one or two types of this behavior, and at different life stages, or working with dogs of another type, dogs will change behaviors. Although young Tucker did not kill coyotes, he eventually pair-bonded with FloJo, and this pair became veritable rangeland murderers for coyotes. They became so effective in hunting predators that the large range flocks they protected did not suffer any depredations. It was this pair that we saw take a small herd of sheep out to graze during the day, and then hide nearby to ambush approaching coyotes.

The strategies used by the dogs are somewhat complex and vary depending upon the situation. Some guardian dogs will attempt to move the sheep away from danger. For instance, instead of attacking a bear entering the pasture one day, I saw an Akbash female gather the herd and chase it to the other end of the pasture. A few years later, this same Akbash actively challenged and hazed bears away from the herd in a large open pasture.

We've seen some of our guardians hide near dead animal carcasses, waiting for predators to approach so they could attack. It's as though they were "baiting" their foes. In another case, our small herding dogs would alert the guardians to the presence of a predator, and would initiate chase, only to have the guardian dogs surge forward and kill the coyotes.

Livestock Guardian Dogs

Some guardians follow the sheep during the day as they graze, but other guardians lead the herd in the day's activities. Most guardians actively patrol during the night, and sleep all day amid the sheep or nearby where the sheep are in view. A study of Maremma guardians in Australia found that the guardians are most active in the early morning and late afternoon, as well as maintaining a relatively high level of activity during the night—the same activity pattern as predators in the area.[11]

When a stranger comes to our home or rangeland to visit with us, the guardian dogs will investigate, always staying between the flock and the stranger. Sometimes the dogs will appear friendly, leaning their bodies against the legs of the stranger, a ploy that keeps the stranger occupied by the guardian.

Most range sheep operations in the United States use four to five guardian dogs with each flock. The dogs work well together in facing predators, with reaction ranging from aggressive barking to full attack. In some cases one dog will react, and other times, all the dogs will be involved.

Guardian dogs are independent decision-makers, judging each threat and reacting accordingly. They generally do not take commands beyond a few basic ones (with the exception of some obedience-trained dogs mentioned previously). Although different breeds may have overall behavioral types, there is an abundance of individual variation within each breed.

The guardians patrol around the herd, with both males and females marking the area of patrol with urine and feces, as well as by barking. Range dogs don't simply guard a territory—they guard the territory their livestock is using at the time. Researchers in Australia noted that regard-less of the size of the area the livestock use, guardians are likely to have a larger range than the livestock they guard—even to the extent that the range includes other livestock on neighboring range.[11] That's something we've repeatedly experienced over the years with a variety of guardian dogs.

Guardians will bark at different occasions, for numerous reasons, and with various intensities. One type of bark appears to simply be an announcement that the territory is occupied, while others are alarm barks, some with excitement or aggression, while others are communications with distant animals. Shepherds who live with the dogs learn the distinctness of these barks, and react accordingly.

Within the last few years, there has been increased debate and questioning about the role of size in livestock protection dogs, as well as the proper number of dogs to run with various-sized flocks of sheep and goats.

Any consideration of either dog size or numbers should be based on the specific circumstance of the flock. What are the primary predator or predators that threaten the herd, and how constant is this threat? Is the herd a stationary farm flock, or a migratory range operation? Do the sheep flock together well, or do they scatter? Are coyotes and domestic dogs the primary threats, or are wolves and grizzly bears? What are the owner's expectations of the dogs?

We've found that the mere presence of a few loudly barking dogs declaring their territory keeps most wild cats, particularly mountain lions and bobcats, at bay. The only time we've had mountain lions successfully kill sheep on our ranch is when a small bunch of lambs wandered off while grazing the lush growth inside an irrigation ditch, away from the safety of the main flock. We

A pair of guardian dogs works well together to protect a small stationary flock where the primary predator is the coyote.

found the eleven lamb carcasses where the mountain lion had executed a sneak attack, swatting lambs up and down the bank, carrying one away for a feast, but killing them all in the getaway. For the most part, when it comes to members of the big cat family, Jim and I have not seen that the size of the dog makes a difference. The dog's attentiveness to the flock, and its constant presence within the flock, are the more important factors. In our area, where mountain lions are hunted with hounds for sport, these cats seem to avoid guardian dogs.

Bears can seem indifferent to the presence of guardian dogs, but we've found that the size of the dog hasn't been consequential in encounters with bears. Our rangeland neighbors have constant problems with grizzly bears in their flocks while on summer pastures in the mountains, and their Great Pyrenees guardians team up (two or three dogs at a time) to harass bears. Our friends in Bulgaria use Karakachans as very effective guardians against brown bears in heavy timber. Although this breed is smaller than other guardian breeds, they can't be beat for pure athleticism, and work as a pack to move dangerous predators away from their flocks.

Wolves pose an entirely different threat to flocks. They tend to view guardian dogs as competitors for territory, or as potential mates (this is rare, but it happens in low-density wolf populations when other mates may be hard to find). Because of these fac-

A mixture of young dogs with older, more experienced dogs is recommended where the predator challenge is varied.

tors, some believe that guardian dogs can serve to attract wolves, but I believe that more often, wolves simply weigh the risk involved in obtaining a feast, and guardian dogs are part of the equation.

Wolves are highly intelligent predators and they can count—meaning a wolf or a pack of wolves will be able to ascertain whether the guardian barks echoing through the night originate with one or two guardian dogs, or a larger dog pack. Wolves tend to take the path of least resistance to a meal, so unguarded herds are easy pickings. A flock with one or two dogs generally doesn't pose much of an obstacle to a pack of wolves, but a larger pack of dogs might be avoided. We've learned that when wolves kill guardian dogs, the wolves usually both outnumber and outweigh the dogs.

After wolves killed several of the dogs we raised, we added Central Asian Ovchar-kas to our guardian dog program. These dogs are generally bigger than the dogs we've raised in the past, and are known as "wolf-wrestlers" in their countries of origin. Spanish Mastiffs are also known to be effective against wolves and are one of the larger guardian dog breeds.

Wyoming goat rancher Carolina Noya notes that the number of dogs protecting a flock is a personal decision with economic ramifications. While a large pack of dogs may be needed in wolf and bear country, there is a financial decision to be made as well. With 1,200 nanny goats, a pack of twelve guardian dogs isn't economically feasible in her situation, with the feed bill for the dogs more than the kid crop's profit. She said, "It is cheaper to feed a few predators than to feed twelve dogs."

Bigger isn't necessarily better, and sometimes moderation is key. For a migratory

Athletic and powerful, a charging Central Asian Ovcharka's muscles ripple as she moves. This female dog, along with a male sibling, worked together as yearlings to maul a black bear that entered the flock. The bear did not survive.

range flock, really big dogs may not be able to travel well, and as a natural result of the way they live, range dogs tend not to be giants. Instead, their days on the range trail means the dogs are very fit.

Macedonia's Shar Planinetz or Šarplaninac (Shar) is a slightly smaller guardian dog breed, but its heavy-boned structure and its canine aggression make this a viable breed for protecting flocks from wolves.

Some guardians will realize the danger of an approaching pack of wolves and will retreat. Others will fight to the death. I've known guardian dogs that have success-

fully fought off wolf attacks, and that have been successful at killing wolves. None of these dogs were small in stature.

Moderate-sized dogs can be fast, and this can be a handy characteristic when your pesky predator is a fox or coyote. Bigger dogs can't run down and kill coyotes or fox, but smaller and younger dogs generally can and do. Our young Akbash dogs are fast enough to kill coyotes, but my chubby, nine-year old female Akbash is not. A migratory range dog in her youth, she's retired to a fairly stationary flock, and as each year passes, she gets wider. Instead of killing coyotes that come near, she simply

Producers must weigh their flock's level of protection to the predators present.

doesn't let the predators enter her flock, and lets the younger, faster dogs initiate the chase.

On our ranch, we've found our guardian program works best with a mixture of dogs of varied ages. Each has its own skills and personality, with a few serving as a constant and calm presence in the middle of the herd, others patrolling out and about, and others prone to pursue predators across the rangelands, a viable option on our open rangeland but perhaps not in a stationary flock. Young dogs learn from the older dogs, and the varied skill levels provide protection from the variety of challenges they face.

When acquiring new guardian dogs, producers should be sure that the dogs are from working lineages, a considera-

tion that is more important than size and number. There are magnificent "champion" dog lineages, but in reality there is no competition for judging potential effectiveness at guardian duties other than the results realized on the ground with livestock. That doesn't mean that champion dogs can't be good guardians—but the more important characteristic is that they are good guardians.

When dealing with live animals, there are no concrete rules or formulas. Just because one combination works for one operation doesn't mean it will work for another—each producer knows his or her situation, and it's best to build a guardian dog program that best suits individual needs. Achieving the right balance in a

Central Asian Ovcharkas can help to tip the balance in favor of livestock protection in wolf country.

guardian dog program can take time and experimentation, and the situation on the ground is ever changing, requiring adaptions on our part as well.

The one undisputable fact is that with effective guardians tending to a flock, losses to predators will be reduced. Producers should worry less about size and num-bers, and more about quality, and what suits their particular situation.

Coppinger and co-authors wrote some good advice in a 1988 paper: "The natural variation in guarding dogs can be capitalized on by matching its behavior with the type of livestock operation and/or the style of the grower."[5]

An Akbash with a red fox she has chased into a culvert in the sheep pasture. Fox can be deadly predators on newborn lambs.

Guardian Behaviors

CHAPTER 6

COMPLEX RELATIONSHIPS

An attentive guardian dog always wants the company of its flock.

The bonding and socialization process works both ways—the sheep must be socialized to the dogs. A sheep flock that is not accustomed to guardian dogs will run when a large, strange dog approaches, and this can cause a chase reflex in the dog (especially in young dogs) that must be corrected by the shepherd. The socialization process can take days, weeks, or months.

The sheep are not mere pawns in the guardian relationship. The sheep and their guardians have individual relationships.

When we bring in a new, but experienced guardian dog, we walk the dog on a leash to the flock for an introduction. Without fail, the dog stands attentively, tail slowly wagging, sniffing and greeting each sheep that approaches to investigate. Within minutes we can generally set the new dog free into the herd, where introductions continue. The body language of both the dogs and individual sheep is mesmerizing to watch, as this cross-species communication continues. Some of the guardian dogs are very physically affectionate, licking ewe noses and lamb butts, while others are more restrained. The more aloof dogs seem to require a bigger personal space, preferring to simply guard rather than actively interact with their charges.

The same can be said of the sheep. Some of the sheep will approach the dogs for affection and attention, while others stamp their front hooves, demanding the guardians keep their distance. And of course some of the sheep show affection for particular dogs, but not others. These are all individual relationships, and any new sheep that is brought into the flock must be examined and introduced to both the dogs and other sheep as well. Both species of animals quickly recognize individuals of the other species. One sheep is not the same as another, and likewise with the dogs.

Although it is true that the guardian dogs "become members" of their flocks, the dogs do not have a mistaken belief that they are sheep. Claims that the dogs believe they are sheep fail to recognize complex animal social relationships, or canine cognition (such as awareness, perception, reasoning, and judgment).

A guardian dog newly introduced to a flock greets the ewes and lambs with calm interest.

Our ewes give birth out in the sagebrush of large pastures—the smallest about one square-mile in size. The guardians move freely among and around the flock, and when a ewe goes into labor, she usually moves away from the main herd, often accompanied by one of her "buddies"—another ewe that hasn't yet given birth. As a guardian patrols and finds a ewe beginning labor, the dog will recline nearby, usually turning its body away from the ewe and looking away. The dogs rarely look directly at the ewe, and I suppose they do this in order to appear non-threatening to the vulnerable animal being guarded. This behavior shows some demonstration of self-awareness.

A guardian stays close to a ewe as she gives birth to twin lambs, providing protection but not looking directly at the ewe.

When I patrol the pastures at sunrise, I first check the main flock, and then drive around looking for each of the guardians, not stopping until I find each one, because I understand the absence of a guardian usually means that the dog is watching over a ewe or lamb. On one occasion in the spring of 2010, I was looking for Rant, a Central Asian Ovcharka, during my dawn patrol. He arose from his vantage point atop a hill, and upon making eye contact with me, he began wagging his entire hind end in greeting, then swung his head and shoulders to the left, looking behind him. He then turned to look at me again, again wagging exaggeratedly, and did the same movement to the left. Rant was using body language to communicate with me the need for me to look behind him, to the left. When I approached closer and could see in the direction he had indicated, I found several ewes with their new lambs. Rant was very pleased—whether it was because he had communicated so effectively, or because of the presence of the new lambs, I'm not sure.

Luv's Girl, now a mature Akbash female, routinely communicates in a similar manner, although not in such an exaggerated fashion as Rant. One spring, as I patrolled looking for the guardians during lambing, Luv's Girl was the last guardian I found. As I approached a sagebrush thicket, she stood up so I could see her through the brush, wagging her tail as we made eye contact. I called to her in greeting, but rather than stepping forward to meet me, she pointed her nose to the ground beside her, and then looked back up at me, continuing to wag her tail but refusing to move. When I approached to learn why she wouldn't come to me, she looked back down at the ground next to her. There hidden in the brush I found a newborn lamb that had become separated from

Livestock Guardian Dogs

its mother. When I picked up the lamb, Luv's Girl went quickly on her way, back on patrol and returning to the main sheep flock.

These are just a few examples of a guardian's ability to communicate across species lines, something we see with other domesticated social species and how the guardians interact with them. The shepherd is but one component of the guardian's *morra* (social group).

Brian Hare of Duke University's Canine Cognitive Center suggests that the dogs' ability to use human social cues evolved during domestication and were not inherited from wolves. His 2004 paper *Domestic Dogs Use Humans As Tools* puts forth the hypothesis "that dogs underwent evolution during domestication that affected their social cognition. The evolution not only enhanced the abilities of dogs to attend to interspecific social cues, but also has facilitated their propensity to request help from humans when they encounter an unsolvable problem."[12] In essence, the enhanced social cognitive abilities are a byproduct of domestication. While wolves are cooperative hunters that rely on behavioral cues of fellow pack members as well as prey, the domestication of dogs has provided for their enhanced interspecific social communication.

Individual guardian dogs have distinct relationships with other guardian and herding dogs, with the livestock they protect, and with the humans tending to them. A guardian's *morra* includes members of a variety of species, and their social interactions are complex (and not necessarily

An adult guardian squinting its eyes, or "winking" when herd members make direct eye contact, is demonstrating non-aggressive, calm behavior.

based on a simple dominance hierarchy as some propose). In this complex social system, individual guardians have repeat interactions with individual predators. As Alikhon Latifi wrote about the guardian dogs of Tajikistan: "Wolves live in the same area where sheepdogs live. Dogs are well familiar with 'their' wolves, as wolves are familiar with 'their' dogs."

The same holds true for other wild animals that share the same range as the guardian dogs. For a number of years we leased a ranch with mixed uplands and hay meadows situated along the New Fork River in western Wyoming, and after we harvested the hay, it was stored in a stackyard near the corrals for winter feeding. Each winter morning, I drove into the ranch and entered the stackyard to retrieve the hay that would be fed that day. The local moose population is familiar with stackyards, and as many as seven moose would jump the stackyard fence to eat the stored hay. One winter, several moose cows became increasingly aggressive as I chased them out of the yard with the truck each morning—finally getting to the point that I had to repeatedly fire a shotgun to encourage them to leave before I could safely exit the truck to retrieve the hay. In an attempt to discourage the presence of the moose, I started feeding the guardian dogs inside the stackyard. It worked. The dogs wouldn't let the moose enter the stackyard, and the problem was resolved as the moose moved upriver—that is, all but one moose cow and her six-month-old calf. This cow and calf were never aggressive, and as soon as the cow heard my feed truck coming down the drive, she and her calf would exit the stackyard, retreating into the willows. Apparently the guardian dogs had decided these particular moose with their polite behavior were the exception to the "no moose in the stackyard" rule. The dogs lounged in the stackyard, watching the cow and calf munching on hay, until I arrived, at which point the moose would quietly depart. Guardian dogs are very capable of processing information from all these interactions, and do indeed include a multitude of species in their social groups.

When it comes to discussing dog behavior, much is often made of the dominance theory, in which alpha animals rank at the top of a social hierarchy, but I don't put much credence in the relevance of the concept. Hierarchy in the guardian dog's *morra* rarely appears to be based on dominance, as demonstrated when we humorously watch the smallest dog on the ranch—a twenty-five-pound bearded collie named Abe—ruling over the dog food bowls. He eats slowly, picking out each morsel from the bowl to consume, as the guardian dogs wait nearby. Only after Abe has eaten what he wants and moves away from the food, does one of the guardian dogs (four times the size of the herding dog) approach for a meal. If dominance were the driving behavior, Abe would have been a guard-dog snack long ago. As we travel the world to view guardian dogs at work, we notice plenty of small-bodied older women or children tending to their flocks, with massive guardian dogs by their sides. Dominance is not what motivates these dogs to work in partnership with their shepherd.

Dominance may play a larger role in breeding behavior of working rangeland guardians, as males vie for breeding rights. But at the same time, we've noticed it's the bitch that decides which male actually gets to do the breeding.

The effectiveness of livestock guardian dogs can be largely influenced by the flocking behavior of the sheep herd. Some sheep breeds do not flock well, scattering when they graze, making effective guarding very difficult for guardian dogs on large rangelands. Most range herds in the United States use breeds that have strong flocking instincts, which allows for more effective guarding by the dogs.

We've found that guardian dogs are exceptionally perceptive and sensitive, although they may not have outwards signs of such traits. These dogs notice changes in their environment and will physically place their own bodies between their sheep herd and danger, be it a truck barreling down a road about to hit the sheep flock, or a black bear attempting to attack a herd.

We've had a variety of guardian dog personalities on our ranch. We spent considerable time and effort in gentling the wild,

shy pups Tick and Juel to be able to catch and handle them. Luv's Girl was a young adult Akbash when she came into our lives, and had had limited handling before we got her. We worked with Luv's Girl so she is exceptionally easy for us to catch and handle, and load into trucks and enter buildings, but she was more than ten years old before she would allow a stranger to our ranch to touch or pet her. Her daughter Rena was highly human-socialized (intentionally, since we made her an ambassador dog for school and library visits), and is the most prone to being stolen by people who see a beautiful friendly dog out on the range.

Rena was the runt of her litter, but constantly fighting with her bigger siblings when we separated her from the litter and

Rena and Roo, a young burro, protecting a small flock of lambs.

raised her with a small group of lambs and a young burro, at our ranch headquarters. I spent the next year photographing this group as they got to know each other and grew up together. The result was publication of a nonfiction children's book, *The Guardian Team: On the Job with Rena and Roo*, that was honored as the Book of the Year by the American Farm Bureau Foundation for Agriculture in 2013, among its other awards.

Although Rena is also a working guardian on our ranch, since we human-socialized her well as a youngster, she has enjoyed dozens of public appearances, and has helped to educate thousands of children about livestock protection dogs. She is just as comfortable snoring on a bed in a grand hotel in our state's capital as she is rolling in fresh manure on the range. She appears to relish her stardom, eagerly walking into a room of excited children waiting to greet her with her tail curled high and wagging, touching her nose to those of the children. We keep these appearances brief so we don't stress Rena, but we firmly believe the sessions are an effective tool for educating the public.

A runt no more, at thirty-one inches tall at the shoulder and weighing 130 pounds, Rena is a big, beautiful dog that most children want to hug and pet, and she accepts graciously. We call her "The Queen" because of the royal way she works a crowd. Rena has also charmed those attending numerous agricultural board meetings and events, and even a few government officials from our nation's capitol. Rena's ability to transition between her public appearances, only to hold wolves out of her flock at home on the range, demonstrates the guardian dog's ability to adapt to a variety of conditions and challenges.

It is heartening to read old journals and articles to learn that some animal husbandry practices used today are simply a continuation of practices begun long ago.

J. H. Lyman described how Spanish Merinos were used in the early American sheep industry in an 1845 *American Agriculturist* article. He wrote that instead of using herding dogs to drive the herd—"a practice entirely unknown"—when a shepherd wishes to move his sheep, he

> calls to him a tame wether [castrated ram] accustomed to feed from his hands. The favorite, however distant, obeys his calls and the rest follow. One or more of the dogs with large collars armed with spikes, in order to protect them from the wolves, precede the flock, others skirt it on each side, and some bring up the rear. If a sheep be ill or lame, or lag behind unobserved by the shepherds, they stay with it and defend it until some one returns in search of it.

For most of our days grazing on the range, this is how we move with the sheep, reserving the use of herding dogs for more crucial movements such as long drives or placing the herd in pens or through obstacles. Our herding dogs are bearded collies, which seem to enjoy being in the company of the sheep, whether called on to move them or not.

Our guardians will keep strange sheep from entering our flock, and every new sheep is inspected. New sheep are recognized as such, and are first investigated by the guardians, and then by the other sheep. Even with our human noses, we know that different sheep flocks have different smells.

Lyman additionally described how the early-day Spanish Mastiffs practiced some of the same behavior we see today:

Sometimes different flocks are brought into the same neighborhood owing to scarcity of grass, when the wonderful instincts of the shepherds' dogs are most beautifully displayed . . . if two flocks approach within a few yards of each other, their respective proprietors will place themselves in the space between them, and as is very naturally the case, if any adventurous sheep should endeavor to cross over to visit her neighbors, her dog protector kindly but firmly leads her back, and it sometimes happens, if many make a rush and succeed in joining the other flock, the dogs under whose charge they are, go over and bring them all out, but, strange to say, under such circumstances they are never opposed by the other dogs. They approach the strange sheep only to prevent their own from leaving the flock, though they offer no assistance in expelling the other sheep. But they never permit sheep not under canine protection, nor dogs not in charge of sheep, to approach them. Even the same dogs which are so freely permitted to enter their flocks in search of their own are driven away with ignominy if they presume to approach them without that laudable object in view."

The problem of bringing new animals into a farm or ranch is addressed in a fascinating way at one farm we visited in Bulgaria in 2010. The large sheep barn has a center enclosure constructed of wood and

A guardian dog shows great interest in a new kitten arriving on the farm, but the first few weeks of such meetings are conducted from the safety of a pet crate or kennel. When this dog became nervous or excited, he would engage in jaw popping—loudly clacking his teeth in a somewhat frightening display.

chicken wire. Any new animal is placed in that enclosure for a period ranging from a few days to a few months, so that all the other animals on the farm are able to view it, visit through the safety of the fence, while the animal is being acclimated and begins to pick up the smell of the farm. This pen was used for everything from chickens and cats, to dogs and sheep. It enables a new animal to gradually transition into its new identity as a member of that farm's *morra*.

In some areas of Africa, new sheep or goats are first rubbed with dung from the local corral or herd before introduction, or penned with members of the native herd for a few days so that when the new animals meet their guardian dogs, they have the smell of that herd.

Rules such as "never move a dog to a new herd or different type of livestock" are meant to be broken. On our ranch, we easily incorporate new dogs into various sheep flocks, and use the same dogs to guard both cattle and sheep. It's also worth noting that although early exposure is recommended to ensure the animals become well bonded, some older individual dogs of guardian breeds can also make good livestock protectors, even if the bond is developed later in the dog's life. We've learned to never say never when it comes to guardians.

Guardian dogs have a different relationship with herding dogs than with other guardians—even herding dogs they have grown up with. Guardian dogs know and understand the differences between guardi-

Guardian dogs interact with herding dogs, most often without conflict.

ans and herding dogs and their duties. Luv's Girl, our most calm and steady guardian, shows affection to her sheep and to other guardians, and even the herding dogs when the dogs are not working. But when a herding dog begins to work her flock, her disapproval is evident. She will run to knock down a herding dog bothering her sheep, and will stand in the entrance to the pen into which the herding dog is trying to move the flock, blocking the herd from entering. When we ship lambs to market in the fall, Luv's Girl whines at the door to the trailer, and runs behind the departing trailer crying her distress as the lambs are shipped away.

Those relationships and resulting behaviors can vary by the situation, as evidenced by the way guardians interact with the herding dogs associated with their *morra*.

Our guardians grow up interacting with our herding dogs, and they know them well. The guardian dogs are friendly to the herding dogs, and often involve them in playful interactions. Our herding dogs never harass the sheep, and they have daily contact with the flock, outside of "working" the flock. The sheep are able to judge the behavior of the herding dogs, and know the difference between a herding dog that is just "visiting" with the flock, and when that dog is approaching the flock in order to work. The guardian dogs are well aware of this difference as well, so there is rarely any conflict.

When we gather the sheep for vaccinations, the guardians go through the processing chute with their sheep, and at shearing, at least one guardian insists on being inside the pen receiving the freshly shorn sheep.

A guardian dog serves as a calming influence for the flock on shearing day.

The dog's presence is welcome, and seems to calm the sheep in a fairly stressful situation.

In the winter, the herding dogs bunch the sheep and move them away from the hay stackyard each morning so I can open the gate, back the truck in to retrieve the day's feed, exit, and close the gate behind me. The guardian dogs know the routine, and move with the sheep under the direction of the herding dogs, without any problem.

A guardian dog's devotion to its flock can seem somewhat extreme. We know of numerous cases of dogs staying with a few sheep alone for months because the sheep had become separated from the larger flock. The fall of 2012 provides an example, and involved a young guardian that had little handling by humans.

I first heard about the dog when I walked into the house one cold, snowy evening and saw the blinking light on the answering machine. An elk hunter had seen three domestic sheep with a white guardian dog up in the Wyoming Range Mountains of the Bridger-Teton National Forest. With range herds leaving the mountains more than a month before, the dog was alone with its sheep. We'd already heard the stories that trickled off the mountain about how hard range sheep herds were hit by wolves that summer and fall, by a pack of wolves and even grizzly bears that had recently expanded their range to encompass the Wyoming Range. More than two hundred sheep had been killed, and the flocks scattered under the pressure of attack after attack.

Several more calls came in the following weeks, but the sheep and their dog were always on the move, never seen in the same spot but gradually making their way from higher elevations as the snow pushed them to lower ground. What sustenance the dog survived on is unclear, but we assume that she caught hares, jackrabbits, and perhaps some mountain grouse. At some point, she lost one of the sheep, and by the time she arrived on ranchlands along the Wyoming Range foothills, she was busily protecting two ewes.

Ranchers and residents of a small rural subdivision saw the devoted guardian dog and her two ovine companions and began watching for them. A few dropped food for the dog when the opportunity arose, but she was shy and wouldn't come to any human.

One day Bill was riding his four-wheeler on a rural road and came upon the frantic dog—one of her sheep had tried to go over a cattleguard and had fallen through. The sheep was firmly stuck, legs dangling down through the grate. Bill enlisted the help of another person, and with the dog barking and growling at them nervously, they hefted the ewe from the cattleguard, freeing the animal. The dog immediately switched from her aggressive, worried barking to rejoicing yelps. She jumped up and down into the air with glee, and then dove down at the sheep's front legs, licking and "kissing" the ewe from below.

Admiration for the dog increased as word of the dog's devotion to her small flock spread, and Gudrid, a local ranch woman, began trying to approach the animal, laying out hay for the sheep and food for the dog until they began staying near her ranch headquarters. The owner of the sheep and dogs was located, but when he arrived, he could not catch the dog. He loaded up the ewes and eventually gave up on the dog. He lived more than one hundred miles away, and Gudrid agreed that rather than shooting the wild canine, she would try to get her caught.

Gudrid and Holly, another neighbor, called us for help. Jim and I loaded up Rena,

Livestock Guardian Dogs

along with a friendly adult ewe, and pulled into the ranch lot. Gudrid called out to the dog, and she soon appeared from the woods along the river bottom where she roamed without her sheep. The young female Akbash was so relieved to see a sheep that she raced to the ewe, and dropped to the ewe's front legs in submission, licking and nuzzling the ewe from her prone position on the ground. We tried, but failed, to lure the dog into our stock trailer with the ewe. She approached the trailer, circling it, but refused to enter. She played with Rena, but eventually retreated into the trees when we pressured her, trying to get close enough to drop a rope over her neck. We gave up, but returned a few days later to try again, once more failing in our best attempts.

Patience is a virtue, and Gudrid was determined to catch the dog before her family moved south for the winter. She continued to feed the dog, and eventually baited the animal into a stock trailer where she slammed the door shut, finally succeeding.

Gudrid backed her trailer into the gate of my dog kennel later that morning, depositing the shy, wild guardian to the safety of captivity. I had let a few ewes into the kennel with the dog, and she soon calmed, comforted by the companionship of the sheep.

It was only a matter of hours before we could come close enough to the dog to drop food and treats. A few days of close captivity calmed her down, but when our sheep flock walked past the kennel the dog cried, wanting out of confinement to join

Introducing new animals from the safety of a kennel is recommended. When she arrived at our ranch, Kit spent a few days in our kennel with some gentle ewes.

Kit meets another livestock guardian dog, Luv's Girl, as they begin sharing duties guarding the flock.

the herd. She met our other guardian dogs through the safety of the kennel panels, and I took Hud, my friendly herding dog, into the kennel to associate with the guardian, and saw she wasn't aggressive to other dogs. A few days later, she escaped the kennel and immediately began guarding our flock.

With her regained freedom, the dog became braver. When she saw the sheep approach me to dig through my coat pockets for treats, she became curious, and was soon eating out of my hand, although she avoided my touch. Jim cut hot dogs into small pieces and soon learned that she had a hankering for these tasty morsels. She even came through our open living room door to retrieve pieces Jim had tossed on the floor as a challenge. She began following us as we walked around the flock, allowing more touch.

I sprawled out on the ground one day, which seems to be an irresistible attractant to our dogs. I soon had guardians licking my head, rolling on top of me, tugging on my jacket sleeves, and generally harassing me with comicality. Kit, as we named our visiting guardian, came close, lying down within a few feet of me, mouth open and tongue lolling, apparently finding humor in my antics. She had finally decided that I was no threat.

We quickly progressed from there. As Jim and I walked near the herd, Kit would approach closely, turning her back to us, presenting her butt as an invitation to scratch the base of her tail. This became routine.

Livestock Guardian Dogs

One day as Kit watched my antics with the other guardians, she approached me, dropping to the ground at my feet, lifting a back leg to show me her belly, a sign of submission. I knelt down and began scratching her belly, chest and back, not holding her down or overpowering her, but petting whatever part of her body she allowed. Soon this also became an infrequent routine, always at the dog's initiation. The wild dog simply hadn't had much human contact, and when allowed to define the terms, welcomed getting to know her humans.

From the start, we had no intention of keeping this wild guardian, and we had a friend in Montana, Annabel, who was looking for a younger guardian dog to help out her aging female guardian in tending to her flock of sheep. Annabel volunteered to take Kit when the time was right. We kept Kit for six months, starting as the wild guardian that had never been touched, and our admiration only grew as we watched her devotion to our sheep flock that spring during lambing season. She was a well-behaved guardian, always attentive to the flock, friendly to the other dogs, and she had quickly learned her name, and responded when called.

With the lambing completed and our impending move of the flock to their summer range forty miles away, it was time for Kit to move to her new home as well. I had begun calling Kit to come inside the kennel to be fed, and on the morning of the big move, locked her inside. Jim dropped a lariat loop over her head to catch her, and I slipped a collar and leash on her neck. We

Kit checks progress on a ewe entering labor.

placed Kit into a large dog crate in the back of our pickup truck, and placed two bottle-fed lambs into a crate facing her, so she had the comfort of her sheep for the three-hour drive to Riverton where we would meet Annabel. Later that evening, Kit was safely in Annabel's barn with the lambs, taking food morsels from Annabel's hand as she sat feeding the lambs their bottles. It was only a few hours later that Kit stepped outside and began guarding her new sheep flock, in the valley between the Beartooth Moun-tains and the Yellowstone River, where she remains today.

Kit's not the first dog we know that stayed behind with a few sheep that became separated from their herd. One year, after being part of a pack of guardian dogs pro-tecting a large range flock in the Wind Rivers, Juel went missing. A small group of sheep had become separated, and Juel stayed with these sheep on the mountain when the big herd exited the area in the late summer. She whelped during that time, and although only one pup survived, she tucked that pup into the ribcage of a dead elk. The pup was well protected with easy access to food when a hunter encountered the dogs later that fall. Juel couldn't leave the area because of the small pup, and was later res-cued with her offspring and eventually reu-nited with her flock.

Some of our guardians tend to exercise somewhat primitive behaviors, including regurgitating food for their pups, much as a wolf bitch will do for her pups. The dogs also kill jackrabbits and bring the carcasses back to the puppy den, even though the pups always have access to dry dog food. Other common shared behavior of wolves and guardian dogs include digging dens to whelp; moving pups to a second site after whelping (called a rendezvous site); pup food-begging; food-caching; urine, feces, and scrape marking their territories; and scent rolling.

One late August morning, while I was camped with my sheep flock in the foothills of the Wind River Mountains, I watched an animal come into the pasture from the southeast. As it got closer, I could see that it was a guardian dog. I did a quick count and realized all my own dogs were already in the pasture, so this was a new animal. I drove over to the stock trailer and set out two bowls of dog food, whistling and call-ing to the dog. It was a slick-haired lactat-ing female, wearing a collar. She was very hungry, and was beginning her second bowl of dog food when I left and went about my business with the sheep.

The lactating female dog stayed away for a day, and I was pleased when I saw her making her way toward my camp the next evening. She sat politely at a distance, and let me approach her with a bowl of dog food. Before I sat it down, I reached out and pet-ted her neck and head, then left her to eat. I sat down on the ground a few feet away while she ate, talking to her, and reaching out every now and then to stroke her neck or back. When she had finished eating most of the bowl, she got up and moved a few feet away, lounging comfortably. I walked over to pet her again, and she exposed her belly to me. I gently stroked her, and checked to learn that indeed, she was producing milk. The pups were out there somewhere. She left a few minutes later, and was quickly out of sight.

The next evening, as we swung back by the stock trailer on our pasture patrol, I spotted the lactating female dog, once again hanging out and waiting for dog food. Jim and I fed her and stood by while she ate. She quickly headed out, and we jumped in the truck to follow, hoping to be able to locate her pups. When she started up a ridge out of sight, Jim trailed her on foot, and quickly found two pups in a rock outcrop, but the bitch had continued traveling over the next ridge. There were scattered bones where the pups had been fed. Jim bundled the two pups in his jacket and I reached him with the pickup truck. We dumped the two pups on the front seat of the truck with our herding dog Hud, and continued out on foot to follow the bitch. About a mile farther out, in a rock outcrop on the side of a steep canyon, the bitch took a defensive stance, barking at us, but we could see a pup's head sticking out from the rocks nearby. We climbed down and dug two more pups out of their den in the rocks. The bitch mostly stalked around below us, vocalizing her unhappiness, and crying, wanting to regurgitate for the pups we were stealing.

After we safely had those pups wrapped up in another jacket, I caught the female, but she slipped her collar when I tried to lead her away. She retreated into the canyon while we began the hike back to the truck with the pups, which we guessed were about six weeks old. We locked all the pups in the stock trailer parked in the pasture near my sheep flock. When we set out a watered-down bowl of dog food, they gorged. All of the pups were in excellent condition, but were distressed over the capture and the new environment of the stock trailer.

We left the pups in the trailer overnight, and camped back up the meadow. Around 4 a.m., Jim headed to the stock trailer to make coffee, and found the bitch waiting nearby. He opened the door to the trailer, and the mother promptly joined her pups. We later learned that the bitch belonged on a sheep ranch about seven miles away, but since all of this rangeland is her home, with

sheep flocks traveling back and forth, she had selected a den and whelping site in a rugged part of her range. We deposited the female and her pups back at the ranch head-quarters. Herders in Spain had told us that it is because of the type of behavior that we had just witnessed that pregnant female guardian dogs wear bells on their collars. This way, they can be found when they whelp out on the range.

Many of our young female dogs help tend to unrelated pups that grow up on our ranch. Juel tended to any young animal presented to her, be it lamb or pup. From taking turns babysitting young pups while the mother is out on patrol, to supervising the pups in their first outings with the herd, young females provide valuable lessons to the members of their dog packs.

Vega (a Central Asian Ovcharka) tended to scare the herders when she whelped. One year she gave birth under a shepherd's wagon, and when the herder approached his camp, she barked and growled aggressively at him until he retreated. Despite her barking antics of protest, Pete crawled under the camp to get the pups out, and put them into the cab of his truck, with Vega quickly leaping inside. He drove the small family to our house and we placed the group in my quiet kennel. Pete and I both know Vega well enough to know she wouldn't hurt us, despite all her protest. Within about a week of having pups, Vega's aggression calmed and she was back to being her friendly self.

The spring after her pups were weaned, Pete decided it was time for Vega to return to work from her maternity leave. He loaded her into the truck and dropped her into a herd of yearling ewes that were to join the other dogs guarding that flock, and proceeded about his day. But Vega had decided she didn't agree with Pete's choice of flock

Good-natured, but independent-minded Vega.

and set out on a ten-mile trek. It was only a few hours later that Pete drove in to check a flock of pregnant ewes to find Vega just arriving as well, intent on staying with the flock of her preference—which was also the herd most vulnerable to predation. When she saw Pete, Vega did an open-mouth, lolling tongue grin at him, wagging her butt enthusiastically. Pete laughed in response, knowing the dog was determined to do things her way.

Vega and her sister Helga developed an interesting strategy for killing coyotes. One of the dogs would stay still, crouched low to the ground while the other gave chase, running the coyote back to the awaiting dog. The herders tending to the flocks often cheered this boisterous sister team for their skills.

Livestock Guardian Dogs

CHAPTER 7

ASIATIC SHEPHERD DOGS

Aralbai's red guardian dog at the family *ger* in the Altai region of Mongolia.

I liked Olgii as soon as I saw it from the air. Situated along a braided river corridor in a mountain valley in western Mongolia, its nearly thirty thousand inhabitants are scattered among adobe houses and surrounded by traditional round felt houses called *gers*, or yurts. Driving into town from the air-port, as we swerved around yaks walking untended down the road or being driven in small herds by men on horseback, I caught glimpses of large red guardian dogs on chains in *ger* yards. I wanted to see the red dogs up close, but never got a chance. They were mastiff-sized and red like some of the

– 73 –

chow-chows I'd seen in the United States. They were always chained up and never roaming freely. I'd already been warned that tethered dogs were vicious dogs.

Olgii is about fifty miles from the Russian border and is about the same distance to the Chinese border. This western province of Mongolia remains home to a population that is 90 percent Kazakh. The Kazakhs had lost this portion of their native land to the Mongols when the formidable warrior Chinggis (Genghis) Khan rode across the Asian steppe expanding his empire. I visited just as the winter season began in October 2008, traveling to the country in order to see and meet Kazakh men who hunt using golden eagles instead of rifles, harvesting animals for their fur. The eagle hunters are subsistence livestock producers year-round, but in the cold of winter they additionally become hunters armed with aerial wolves. Viewing the local dogs would be a bonus of our travels.

My friend Janell and I spent one night in a tourist *ger* in Olgii. I listened for most of the night as the dogs from throughout the city loudly barked to be heard. They seemed to effectively communicate, each claiming and defining its territory, and responding to each other through specific barks and howls. At breakfast the next morning, I mentioned to Erlan, our local guide, that I'd listened to the nocturnal vocalizations of Olgii's dogs. Apparently it's such a familiar sound that it goes unnoticed by village inhabitants. Janell noted that in the urban areas of United States, dog owners can be penalized for having a barking dog, which Erlan found humorous as it's such a constant background noise in his hometown.

As we traveled the countryside on horseback, our nightly stops would be at a *ger*, staying as guests to a family, and some-times with pastoral herders. The standard greeting when approaching a homestead in Mongolia is to shout out, "Hold your dog," as every *ger* seems to have a guardian dog, and shepherd dogs are found throughout the steppe along with herdsmen, protecting herds of wooly sheep, cashmere goats, and other livestock. On our first night out in the countryside, we stayed with Aralbai and his family. Their *ger* was guarded by a red dog, which I learned does not go out with the livestock, but guards the *ger* and keeps birds and other scavengers away.

This part of Mongolia felt so much like home to me, and I suppose it's because there are so many similarities to western Wyoming. The Bayan-Olgii province, in which the eagle hunters are located, is Mongolia's highest-elevation province. Mongolia's highest peak is the Altai's Tavan Bogd Uul (Five Saints Mountain), which is just over 14,350 feet. Wyoming's highest mountain, Gannett Peak, is nearly as tall, and I see it out my living room window on the ranch.

Livestock herds are grazed on a family's ancestral lands in the Mongolian countryside. Private property is a concept recognized within Mongolian cities, but out in the country, there is really no such thing. Land cannot be bought and sold, and its use is based on ancestral tradition and family relationships. The result is that there are no fences to hinder movement through Mongolia. As we rode, we encountered both *ger* dogs, and dogs with grazing livestock herds. Some of the dogs approached us aggressively, and escorted us past, with hackles raised, and barking in warning.

Michelle Morgan Esterhuizen, a development economist with experience in Mongolia, noted that while it's still possible to see the massive dogs regarded as the "true Mongolian dog" living alongside herders in

As we rode horseback across the countryside, guardian dogs escorted us through their territories.

remote areas of the countryside, most dogs in the remainder of the country have the typical coloration and markings, but are of smaller size or of mixed breeding. Mongolian guardian dogs can either be long-haired or short-haired, and most are black with tan or yellow or white markings on the chest, face, legs, and tip of the tail. It is said that the sign of a true Mongolian dog are its pair of yellow or tan spots above the eyes (called "four eyes").[13]

One night just before dark, we arrived at a larger winter house, complete with separate kitchen and sleeping area, and of course, its own *ger* guard dog. The dog was not friendly, but settled down for our visit. It was a short-haired dog with the four-eyes

trait. We'd seen others as we rode horseback across the steppe.

We drove to Sagsai, the village hosting an eagle festival, and settled into the Blue Wolf Tourist *Ger* Camp there. A tourist *ger* camp usually consists of about a dozen *gers* set up for visitors, with a central *ger* serving as a restaurant where meals are served, and separate buildings for toilets and showers, with a bar adjacent. A security fence encloses the entire site.

Two men guarded the camp and another two were in charge of lighting the fires in the *gers* at night. We had been warned not to touch the vicious black and white *ger* dog that guarded the camp. Unable to resist the challenge, I reached toward the dog's head and

A short-haired "true Mongolian dog" exhibiting the black, tan, and white coloration and pair of tan spots above its eyes.

he snapped at me in threat. Since that failed, I reached over and scratched the base of his tail. Within seconds, he was leaning his butt against my leg, growling and snarling at anyone who came near me. The *ger* guard dog adopted me, guarding our *ger* for the duration of our stay. Janell and I laughed as we watched him repeatedly chase a man at a run through the camp. I don't know why the dog didn't like the man, but I trusted the dog's judgment. The dog would find me at various locations during my time there, always escorting me safely back to my *ger*, growling and snarling at anyone approaching too closely, always keeping his body near my legs.

There are disputes about what to call the various types of large guardians found in Mongolia, but the most common names seem to be Knonch Nokhio, Bankhar, and Mongolian Shepherds. We saw a variety of herd guardians in our horseback journey, and met one battle-scarred male dog alone on the steppe. An American nonprofit organization, the Mongolian Bankhar Dog Project, is spearheading an effort to retain and restore the traditional guardian dogs to herders across the Mongolian steppe.

Many of the ancient livestock guardian dog breeds in Asia, although used for thousands of years, have neither been studied nor written about extensively in English-language journals, so discovering information about them can be somewhat difficult. There are numerous papers in

A dog known for its viciousness at a *ger* camp that opted to protect the author from disturbance.

scientific journals about the effective use of guardians around the globe, but dogs in Asia are usually not included. Fortunately, some of the very best information about guardian dogs that can be used by livestock producers is found in the journals of the International Society for Preservation of Primitive Aboriginal Dogs in its Russian branch publication, which is now an international journal. These journals are written by people who live in close association with dogs around the world, and livestock guardians are a group that is frequently addressed, specifically their historic use worldwide.

Russian scientists described several variations of guardian dogs in the Tuva region of Siberia, where Mongolia and Russia meet. The Tuva Ovcharka is described as having three variations, including a very big dog used to guard camps; a black mastiff-type dog used to guard herds that are often called four-eyed dogs due to red spots above the eyes; and a lighter hunting dog. Many of these Tuva dogs were lost with the forced relocation of nomadic peoples to villages under the regimes of the Soviet Union, which also implemented dog extermination policies. Fortunately, some of the animals survived and can still be found on remote grazing lands. These Tuva Ovcharkas are similar to Central Asian Ovcharkas, but do not achieve the large size found in Ovcharkas in other regions.

A typical herd guardian found on the steppes of Mongolia.

Some believe that the origin of the Tuva Ovcharka is intertwined with that of the yak traditions of Tibet. Dogs guarding live-stock came together with aboriginal dogs guarding Tibet's yak herds, and served as the ancestors of what we know today as the Tibetan Mastiff. The breeding of yaks came to Mongolia from Tibet, and with the tradi-tion came the guardian dogs.

The most common guardian dog present throughout Central and Southwestern Asia is the Central Asian Ovcharka, a breed that is known by a variety of different names based more on location than morphological dif-ferences. These guardians are believed to be one of the oldest breeds of dogs on earth and were raised in their countries of origin from birth with sheep, in the traditional nomadic way of life common to subsistence livestock production. The dogs guarded flocks and their people, and dogs that showed aggres-sion toward humans were destroyed.

Alikhon Latifi and Arunas Derus reported on the decline of these dogs over decades, noting that in the 1960s, Central Asian Ovcharkas were the only dog breed

A battle-scarred dog that the author encountered on the Mongolian steppe.

found throughout the Tajikistan landscape. But by the late 1970s, other dog breeds were introduced, and a few popular movies increased the Ovcharka's popularity, resulting in people bringing the dogs into cities and villages. Newly-introduced diseases took a toll on the dog population, and the eradication of the wolf population made the need for working guardian dogs obsolete. Dog fighting enthusiasts began tapping into the remaining Ovcharka populations, breeding for fighting characteristics rather than guardian duties. Conditions further deteriorated with the civil war in Tajikistan.

"There is a saying that during wartime, the number of wolves and bad people always increase," Latifi wrote. "Tajikistan experienced this in full."[14]

Dogs in nearby Iran were devastated as well from the eradication of the wolf population. During the Iran–Iraq war, dogs were found consuming human remains, and teams were dispatched to exterminate the dogs—including dogs far away from battlefields.

Derus reported that when he toured the Tajikistan countryside to view guardians out protecting their flocks, he was pleasantly surprised at the number and quality

Asiatic Shepherd Dogs

of the dogs he encountered. He described stopping the car and stepping outside near a flock grazing alongside the road. Two Central Asian Ovcharkas walking behind the herd sped up to take a position between the strangers and the flock.

"It looked like a game of checkers, when our every move is blocked by the opponent," Derus wrote. "As we observed later, other herd guarding dogs were also skillful in using territorial tactics so that we remained constantly cut off from the sheep and we were thus kept in a very inconvenient position, if we wanted to access the herd."[14] Some dogs encircled the men at different stops, and some attacked their car. The men never saw fewer than four guardians with any herd, and flocks were commonly guarded by seven or eight dogs. There were many more males than females, and herd-

ers kept a greater number of dogs in areas where bears prey on their herds.

It is reported that in the mountains of Uzbekistan the dogs are now ever-present in most households rather than in sheep flocks. The dogs guard their territories, and neighboring dogs in a village will guard their general vicinity from strangers. Dogs are treasured members of their families, but dogs that bite people are killed.

There are a number of reasons for the decline of native Ovcharkas of central and southwestern Asia, including the loss of nomadic traditions due to political and social changes and the modern use of Ovcharkas in organized dog fights for money. Fortunately, there are pockets of nomadic peoples throughout Asia that have never lost their traditions of using Central Asian Ovcharkas as loyal guardians.

The Central Asian Ovcharka is believed to be one of the most ancient of the guardian breeds.

CHAPTER 8

LARGE CARNIVORES, NEW CHALLENGES

Sometimes the only trace of the difficult work of the night is a weary, bloodied guardian approaching for food or attention the next morning.

Only the late night and the trees are witnesses of those vicious fights with countless foes, in which this remarkable breed of dog was forged," Zaur Bagiev of Russia wrote about Ovcharka guard-ians in a 2006 issue of the *Journal of the International Society for Preservation of Primitive Aboriginal Dogs*. "It is impossi-ble to count how many of them died from a wolf's canines while defending their human friends and their property. They defended exactly their friends, not masters. Only a fearless dog that did not know slave-like submissiveness and with an unbroken spirit could go into the darkness full of glowing hostile eyes and hungry jaws."[15]

Recovery programs for large carnivores in the United States have been successful, with predator populations expanding both in number and in range. The recovery of gray wolf and grizzly bear populations has been accompanied by increased conflicts with livestock that share the same range. Wolves or grizzlies, or both species, now inhabit areas that were free of large carnivores for decades. This range expansion occurred while these predators were under federal protection granted by the Endangered Species Act, so when depredations on livestock occur, management options are limited. Because of the protected status of both grizzlies and wolves, wildlife managers now seek alternative management options rather than automatic lethal control of these predators when conflicts arise. Predator management efforts are aimed at preventing or reducing the frequency or severity of conflicts, dealing with the individuals that cause the conflict (most often through removal), and increasing local tolerance for carnivores (through education, compensation, harvest, etc.).

Western Wyoming's pattern of land ownership includes the majority of acreage administered by federal agencies, with most of the remainder held in private ownership as ranches. The lowest elevation of this arid region is about 5,500 feet, and large acre-

A herder moves across the Wyoming landscape with his *morra*.

ages are needed to provide enough forage for cattle and sheep operations. Thus, most ranches graze their herds at least a portion of the year on federally administered land. Nearly 70 percent of the nation's sheep inventory resides in the western states, and an estimated 25 to 30 percent of all sheep in the United States graze on public land allotments.

Some range sheep flocks spend nearly all year on public land, grazing in lower-elevation deserts free of deep snows in the winter, moving to higher-elevation mountain pastures as the snow melts in the summer. Sheepherders live in wagons alongside the flocks, with camptenders checking on them every few days and bringing supplies to the herders. Similar to nomadic cultures in other areas of the world, range sheepherd-

ers in the western United States also practice transhumance, the act of moving herds with the seasons and using guardian dogs to protect them. Flocks graze over hundreds of miles of range and often go unnoticed by the public although their grazing practices are regulated by federal agencies. Most range sheep operations utilize land that is primarily unfenced and unimproved, and that involves long-distance movements from season to season and on-site herders.

The guardian dogs live with a flock full time. Most herds consist of highly gregarious western, white-faced sheep, primarily of the Rambouillet breed. They have a strong flocking instinct, which helps in defending against predation. Each *morra* usually consists of about one thousand ewes and their lambs, with two to five protection dogs,

Migratory flocks often spend summer and fall in high-elevation mountain pastures.

and a herder. The flock spreads out up to about one square mile to graze during the day, but beds together in a tighter group at night.

Range-sheep producers in western Wyoming have individual ranch operations, with changing guardian dog populations. For example, one sheepman might place five dogs with each band of sheep, and three bands are trailed to separate mountain pastures. Some bands may retain all their dogs, while other bands may gain or lose dogs, so that one band may have two dogs and another may have seven. In other cases, the dogs associated with each band may not see other guardians until late in the fall, when the sheep come off the mountain and the dogs converge at sorting pens. The dogs belonging to one ranch will travel with their flocks to a shared winter range, where herds and dogs from other ranches are encountered. The guardian dog population constantly changes as the dogs mature, are hurt, or die. Some dogs may leave their herd to go with an adjacent flock, and owners will switch ownership of dogs, "borrowing" male dogs for breeding.

The most common guardian dogs used in the United States are the Great Pyrenees and the Akbash, with the Anatolian Shepherd, the Maremma, and the Shar Planinetz used to a lesser extent. These dogs usually

Migratory livestock herds on western rangelands spend part of the year in lower-elevation desert country.

weigh from seventy-five to one hundred pounds (but can be as large as 130 pounds), and have been very successful at reducing predation from coyotes.

In the United States, guardians are used to protect against coyotes, but guardian dog breeds originated in Asia and Europe to combat predation by a variety of predators—including brown bears and wolves. While many guardians are successful at repelling black bears and grizzly bears during most encounters, their effectiveness against wolves has met with mixed results.

A lone guardian may not pose much of a challenge to a lone wolf. Federal wildlife officials in the United States have even documented a few cases where lone wolves and lone guardians settle into a somewhat companionship relationship, and the wolf was able to kill sheep without the dog's inter-

vention. In other cases, lone wolves killed lone guardian dogs.

One of the primary causes of death of wolves within Yellowstone National Park is from attacks by other wolves—even among fellow pack members. A wolf pack that has established a territory will defend that territory from other canids, including livestock guardian dogs. Livestock producers in these areas will generally need to use more than a few dogs to protect their flocks from wolf depredations. Montana livestock producers using two guardian dogs to guard a sheep flock on rangeland utilized by a pack of six to nine wolves had both of their guardians killed in one event. The shepherd camped with the sheep attempted to intervene to save the dogs, to no avail.

In another case, a pack of wolves from Yellowstone National Park routinely visited

Guardian dogs used to protect a migratory sheep flock in Wyoming.

a Montana sheep ranch and howled at and attempted to fight with the ranch's guardians, killing several in separate incidents. The ranch owner locked up the remaining dog in order to keep it safe, believing the wolves were attracted to the dog for some reason.

Most of these wolf–dog incidents occurred during the time and in an area where the most common livestock guardian breed in use was the Great Pyrenees, which is not one of the more aggressive breeds that, I believe, is more suited to facing wolves. In 2012, rancher Jeff Siddoway wrote about the impact of wolves on the guardian dogs protecting his family's sheep flocks in Idaho, noting: "The guard dogs are the first things the wolves kill. We have lost about eighteen Pyrenees in the last three years. Our herders sleep right next to the sheep at night in teepees. The wolves usually kill between 2 a.m. and 3 a.m."[16] Livestock producers utilizing public lands, like Jeff Siddoway, have to weigh a variety of factors in selecting what breed of dog to use with their flocks. The dogs must provide some level of protection to the herds, but can't be too aggressive to human recreationists and other public land users. It's a difficult decision, and one with ramifications.

Ed Bangs, formerly with the U.S. Fish and Wildlife Service, alleges that wolves perceive dogs as "trespassing wolves" which will be actively searched out and attacked. "This territorial behavior is well documented and appears to manifest itself most strongly when the wolves outnumber or outweigh the dogs involved," Bangs and his co-authors wrote. "Perhaps a more evenly matched battle might still occur between multiple livestock guardian dogs and wolves, but with less injury to livestock guardian dogs, although wolf-to-wolf conflict often results in dead wolves."[17]

Robin Rigg reports that guardian dogs have been injured during encounters with bears and wolves in Slovakia.[10] We know the same is true here in America.

Livestock guardian dogs may not entirely prevent wolves from attacking livestock, but their presence can reduce the number of animals killed in an attack, and does appear to inhibit surplus killing behavior of wolves.

Although guardian dogs are viewed as a nonlethal method of predator control, in reality this is only partially true. In most encounters with predators, the dogs are nonlethal, but in some cases they do kill the challenging predator (most often coyotes; black bears less frequently; and, on occasion, even wolves). But in encounters with wolves, more often it's wolves that do the killing.

In a 2010 issue of *PADS*, Robert Vartanyan of Russia wrote that "once [wolves] have decided to attack sheep, they probe to see whether there are people with the herd and what kind of dogs are protecting it. If the dogs are active, do not hide, firmly keep their positions and give warning that the herd is protected by barking aggressively, then the wolves try to frighten or outsmart them. For example, they howl for a long time: some dogs are scared by wolves howling. They try to distract the attention of the dogs on one flank and attack at the other."[18]

The number of conflicts between guardian dogs and wolves is increasing in the Rocky Mountains of the United States, with eighty-three guardian dogs killed by wolves in this region from 1985 through December 2005, and numerous other attacks since then. Confirmed fatal wolf attacks on guardians are only a fraction of all wolf-caused deaths, since many of the

A Great Pyrenees carries the scars of previous battles.

dogs will simply disappear, with their fate unknown.

Sixty-one percent of the eighteen documented fatal wolf attacks on guardian dogs from 1995 through 2004 in the Yellowstone region of the Northern Rockies involved the killing of Great Pyrenees dogs. These conflicts involved dogs that were outnumbered and outweighed by their wild counterparts.

There are similar reports of wolves killing guardian dogs in France, and both hunting dogs and livestock guardians in Italy. Wolves in the United States have taken a toll on hunting dogs as well. There were forty-nine hunting dogs confirmed as killed by wolves in Wisconsin 2004 through 2006, with an additional ten injured. The number of hunting dogs killed in Wisconsin continues to rise, with more than twenty dogs killed annually, including at least twenty-two dogs killed in 2015. The eighty-six dogs confirmed to have been killed by wolves in Montana, Idaho, and Wyoming during 2006 to 2015 included livestock guardians, hunting dogs, and pets.

In one stunning case in Romania, wolves killed 157 adult livestock guardian dogs from January 2001 to October 2002 in a forty-square-mile area, with wolves consuming the majority of the carcasses in most cases, and leaving the nearby livestock unscathed. This situation appeared to be the result of a lone wolf pack that specialized in preying on and eating dogs.

Large Carnivores, New Challenges

The CanOvis Project: Wolf-Dog-Flock Trinity

Wolf damage to livestock herds in the southern French Alps continues to be a chronic problem, with more than 2,400 head of livestock killed by wolves in 2013. Researchers with the CanOvis Project of the Institute for the Promotion and Research on Guarding Animals have indicated that the region is facing the limit on the efficacy of the use of Livestock Guardian Dogs (LGDs) in that region.[19]

Researchers were able to record nocturnal interactions between wolves and LGDs in the Maritime Alps. Research involved three flocks of sheep, two of which had high wolf pressure, including one grazing in an area where no wolf shooting permits are issued—not even to livestock producers experiencing wolf attacks on their herds. Flock sizes ranged from 1,750 to 2,500 sheep. One area had two flocks at the start of the grazing season, but these were combined at the end of the summer due to frequent wolf predation on one herd. All three flocks were protected by LGDs, mainly by Great Pyrenees dogs, or Great Pyrenees/Maremma crossbreeds. One flock had eleven LGDs, while the other two herds each had four guardians. The LGDs were fitted with GPS collars each evening, and their movements were tracked until sunrise.

How Did the LGDs React?

LGD reactions ranged from no reaction, to barking, social or close contacts (33 percent of the events), and chasing. Using infrared binoculars, researchers were able to document wolves passing by the flock, feeding on freshly killed sheep, and attempting to attack sheep—despite the presence of LGDs. The researchers noted: "Wolves were apparently unafraid of LGDs. Although wolves were chased by LGDs or had agnostic encounters, these experiences did not prevent them from returning the same or following nights. Moreover, we recorded several occurrences in which a single LGD faced a wolf and exaggerated its behaviors instead of attacking, allowing enough time for the wolf to escape. Thus, the LGDs observed (either naïve or experienced with wolf encounters) seemed to be very cautious around wolves."

The researchers suggest that LGDs should be considered a primary repellent by disrupting a predator's behavior, but they do not permanently modify that behavior. Wolves become habituated to the presence of LGDs, according to the researchers. They found that both LGDs and wolves seem to evaluate the risk of escalating confrontation.

Aggression

Great Pyrenees LGDs are often selected for use in areas with high levels of tourism, because they are known to be less aggressive to humans and other dogs. In fact, they are now bred and promoted for their docility. But LGDs that are expected to be effective guardians in wolf territory must have a higher level of aggression to predators. They must have a willingness to confront and fight the predator, as certain LGD breeds are known to do. Researchers pointed to the Karakachan from Bulgaria as a breed known for its aggression to intruders.

Barking
The researchers found that LGD barks do not modify wolves' ongoing behaviors, but these vocalizations do seem to transmit information. "Because barking is easy to pinpoint, they might give valuable information to the wolves about the LGDs' location, the number of individuals, their distance and maybe even temperament. Nevertheless, LGDs' barks can attract other LGDs, even if they are not able to observe the scene."

Young Wolves
Particular wolves were seen staying near the flocks, attempting (and failing) to attack, and interacting with the guardian dogs. Researchers believe these were young wolves learning to hunt and they were testing the dogs.

"Consequently, if these first encounters are not associated with negative consequences, we hypothesize they will learn that LGDs and shepherds are not a danger and will perceive sheep as an available resource," the researchers noted. "This knowledge may then be passed to the next generation through associative learning. Thus, more aggressive LGDs may be necessary to teach young wolves that encounters with LGDs have severe consequences."

Shepherds Aren't a Threat Either
The researchers found that shepherds aren't viewed as much of a threat to the wolves either. Since their only option is yelling and throwing rocks, their effect on wolves is negligible. The researchers found that the wolf flight distance when confronted by the shepherds was sometimes as short as one hundred feet. Recent wolf attacks on sheep flocks are happening more often in daylight (52 percent of all attacks) and a shepherd reported being challenged by a wolf while trying to retrieve a wounded lamb.

For More Information
The CanOvis project is an undertaking headed by Jean-Marc Landry and the Institute for the Promotion and Research on Guarding Animals in Switzerland. For more information, and a series of videos and photographs, check out www.ipra-landry.com.

Because our family raises guardian dogs with our sheep flock, over the years we have sold bonded pups to other sheep producers for use in their flocks. Within a period of about a decade, eight of the dogs we raised were later killed in conflicts with wolves. Wolf pack predation on a flock is not just a traumatic event for the herd involved, but the loss of even one guardian dog is an economic hit to a livestock producer, and leaves the herd vulnerable to further predation. It takes more than a year to raise a pup to be an effective guardian, and the dogs protecting a flock will work as a team, so producers who suffer the loss of a dog to wolf depredation have an immediate crisis that is not easily resolved unless they have another adult guardian in reserve that can be placed with the affected herd.

Jim and I decided we needed to take a look at what other livestock producers do when faced with large carnivore populations on their range. The first part of our research involved conducting a literature review of references from around the world to gain insight into increasing the effectiveness of guardian dogs in wolf range. Our goal was to identify different dog breeds that may be well-suited to facing wolves and could be utilized in the Northern Rockies of the United States where we have increased conflicts with both wolves and grizzly bears.

We defined the criteria we would use in selecting guardian dog breeds that had potential for use in our area, including:

- Must be canine-aggressive, so that the dogs are inclined to actively challenge wolves;

- Must not be human aggressive, since many herds graze on public lands for part of the year;
- Should originate in areas with large carnivores to take advantage of working characteristics similar to that of the Northern Rockies; and
- Must not be of small body size, so that the animals stand a better chance against large carnivores.

Some breeds that may work well in the United States were left off our list because of their rarity and/or the prohibition of export out of their countries of origin, or the fact that the dogs inhabit conflict zones where the possibility of getting access to the dogs would be limited. For other breeds, there simply wasn't much published information available.

As we conducted our research, we learned that just as wolves have been persecuted throughout the world, with wolf populations falling to near extinction in some countries, the livestock guardian dogs that protected flocks from wolf predation had fallen victim to the same fate. Without pressure from large carnivores and with livestock herds decreasing worldwide, the use of guardian dogs had decreased to the extent that there are now recovery programs in place to get these dogs back onto their historic landscapes to address expanding large-carnivore populations.

The spread of communism in Europe and Asia brought with it an active campaign of collectivized agricultural policy, which worked to rid entire regions of its free people—the nomadic livestock cultures. Livestock and their guardian dogs were killed

As wolf populations expand in the Northern Rockies, conflicts with wolves have escalated.

or collectivized, and their nomadic herders and families were taken from the land. When the herders became villagers, the cultures lost their old traditions.

In the 1930s, the demand for dog skins resulted in the destruction of the largest dogs available; rabies campaigns called for wholesale extermination of all dogs in certain regions, including livestock guardians; later inter-mixing of dog breeds resulted in large dogs lacking in guardian instincts; and the introduction of dog diseases all took their toll on native dog populations. The decline in interest in livestock husbandry,

as well as a decline in predator populations as a result of persecution by humans, also played a role in the reduction of working guardians. Recent interest in dogs declared as members of "national" dog breeds has resulted in a greater demand for dogs in the pet trade than as working animals. In addition, some breeds are now being used and bred for dog fighting rather than flock guardians.

Ovcharkas (both Central Asian and Caucasian) are used in dog-fighting rings today, especially in Central Asia and Russia. While there is much (and well deserved) criti-

Two Central Asian Ovcharkas have an uneasy meeting on the range.

cism of dog fighting, traditional dog fights in Asia are not like pit bull dog fights in America, which involve severe injury or the death of the dog. Dog fights involving livestock guardians in their historic context of nomadic people played an entirely different role—that of testing the best dogs as wolf fighters, and promoting the best dogs for breeding. Dogs with the proper drive, tenacity, and strength needed to confront and kill a wolf are selected to pass their genes on to the next generation. Serious injuries are reported to occur rarely since these matches are conducted to observe traits, such as dominance display, agility, and physical strength.

Dog fights, which are called "wrestling" in Central Asia, serve to test a dog's level of canine aggression and have a set of stringent rules developed over centuries. Any dog that is inactive or cries is determined to have lost, and human handlers end the match. Signs of submission end the fight as well. Traditional guardian dog fights or wrestling matches involve the dogs' controlled aggression, not blind fury as seen in pit bull–type fighting.

John Stuart Blackton wrote for *Foreign Policy* in 2009 about his experience watching traditional dog fighting in the wintertime in Afghanistan, noting that the matches were more wrestling than fighting, in which one dog attempts to flip the other onto its back. If both shoulders of one dog touch the snow, the fight is won. If the dogs shift from wrestling to truly biting, the match is ended.

"Like wrestling, the art [is] in leverage, timing and speed," Blackton wrote. "No dogs die. Almost no dogs are injured (they are too valuable)." This type of dog wrestling was a highly anticipated spectator sport, with much cheering and wagering by the gathered crowd.

In traditional guardian dog matches, the fights begin and end quickly, and the result was a determination of the best dogs to fight wolves. One champion fighting dog, a Caucasian Ovcharka that lived in Russia in the 1930s, was believed to have killed one hundred wolves in his lifetime, an enviable record.

Ilgam Gasymzade and Namik Azizov of Azerbaijan wrote that they were eyewitness to the killing of two wolves by a Caucasian Ovcharka wolf-fighting dog in 1977. The willingness to aggressively challenge a wolf can be an important factor in a guardian dog's effectiveness. Bulgaria's Sider Sedefchev told of seeing a Great Pyrenees chase a wolf a short distance before the dog stopped, contining to bark. Sedefchev noted "the dogs show him with their behavior that they are not a real obstacle, and the wolf's success is just a question of time." He continued, "When the dogs chase the wolf with the intention to kill it, this means much more for the wolf."[20]

In contrast to the Great Pyrenees, the Karakachan dog of Bulgaria has been known to chase a wolf away from the flock for nearly a mile and a half. It has been reported that after such encounters, wolves leave that flock alone and turn their attention to other flocks. Janet Dohner wrote in her book *Livestock Guardians*, "The shepherds believe that their dogs must show this level of dedication to harassing and even attacking wolves in order to combat the strong predator pressure in the area; therefore they value physically strong, confident dogs."

The importance of canine aggression in guardian dogs working in wolf territory has become evident in recent research in the French Alps. The CanOvis Project of the Institute for the Promotion and Research on Guardian Animals reported that wolves became habituated to Great Pyrenees dogs when contacts with the dogs did not result in severe consequences (such as physical attacks). Researchers there suggested that protection of the flock "depends primarily on the physical ability of the livestock guardian dog to consistently disrupt predatory behavior night after night or to win a fight."[19] These researchers concluded that selecting for aggression may be beneficial for protection of the herd—a conclusion that we've reached independently based on our own experiences.

The readily available livestock guardian dog breeds that rose to the top of our list

Wolf-Dog Hybridization

Recent research in the Caucasus Mountains of the Republic of Georgia has discovered a surprising amount of hybridization of wolves with livestock guardian dogs in the region. Researchers associated with the Institute of Ecology at Ilia State University and the Tbilisi Zoo (both located in Georgia) found recent hybrid ancestry in about 10 percent of dogs sampled, with two to three percent identified as first-generation hybrids.

as more suited to facing wolves included Central Asian Ovcharka, Transmontano Mastiff, Karakachan, Kangal, and Shar Planinetz.

Jim and I wrote the results of our literature review in a paper published in *Sheep & Goat Research Journal* in early 2010.[21] Our paper recommended that livestock producers in the Northern Rockies try using groups of two to five of these more aggressive guardian dogs to protect each of their flocks, with the objective of outnumbering and out-weighing wolves or other large carnivores encountered. Individual dogs have different behavioral and guarding tendencies, so developing the proper mix of traits into a group of guardians takes time and readjustment of the animals involved.

Those seeking to acquire guardian dogs from their countries of origin should note that although "herding" dogs in the United States are defined as breeds like the border collie, some guardian dogs are called "herding" dogs also, but in a different context. Some strains or lineages of certain breeds are called property protection dogs. These dogs stay within defined property boundaries, and many protect the property from thieves. Other strains of the same breed are called herding dogs because they stay with livestock herds to protect them from wolves and no aggression toward humans is acceptable. The Caucasian Ovcharka is an example of a breed with these two different types of protection dogs, which is also reflected in the physical conformation of

A young guardian dog rests near his flock.

the two. The wolf-fighting or herding type is found in the North Caucasus where there are migratory sheep herds. Zaur Bagiev reported in 2006 that "a pack of dogs resembling a pride of lions is present with each animal herd" and the dogs not only protect the herd, but control its movements as well.[15]

Kazakhstan has the highest density of wolves in the world, and guardian dogs are the only reliable form of protection against predation there. Their importance is reflected in the Kazakh proverb, "The dog is more important than sheep."

Even if livestock producers use the most effective guardian dogs, there will inevitably be conflict between wolves and livestock, and their human and canine guardians. Instead of eliminating this conflict, management programs should aim to reduce the amount and severity of that conflict.

Livestock protection dogs are a deterrent, and they are not a silver-bullet solution to depredations. Wolves will still attack and kill livestock, and usually once a wolf pack starts killing livestock, the only way to halt that behavior is to kill members of that pack—and in some cases, all the members of a pack. We know that some of our livestock will die at the mouth of a wolf, and we know that some wolves will die—from a bullet fired from a rifle, or at the jaws of our guardians. The dogs simply reduce the amount of conflict and damage, but do not eliminate it.

Guardian dogs have been found to reduce sheep depredation by 11 to 100 percent, and most livestock producers surveyed viewed the dogs as an economic asset, with high economic efficiency for a relatively low cost. Livestock producers also note that the use of guardian dogs does not require assistance from government agencies or rely on advanced technology.

In writing about the use of livestock guard dogs in Asia, Tatyana Mikhailovna Ivanova explained the relationship of the nomadic culture, their herds, herd protectors, and the landscape upon which they all depend: "Experience accumulated during hundreds of years is passed on for generations and it has achieved a high degree of perfection that cannot be improved any further but only preserved."[22]

While livestock producers in the Northern Rockies have been able to use some of the same tools used by these nomadic cultures, what has been lacking is a full range of techniques and knowledge of how to use those techniques. Tapping into the knowledge and tradition of ancient livestock protection dog cultures will help to fill that void.

Our paper prompted further interest and discussion about what American livestock producers could do to protect their herds. It also led us on a journey to learn more, as we traveled to Europe and Turkey to see the dogs at work in their homelands.

A Central Asian Ovcharka engages in play with another guardian dog.

CHAPTER 9

OLD WORLD TRADITIONS

A shepherd with his flock in Wyoming.

Although Jim and I are familiar with governmental policies and regulations for livestock producers in the United States, we have learned of surprising differences in these programs in the European Union member nations of Spain and Bulgaria.

After publication of our literature review, we traveled to three Old World regions (Portugal/Spain, Bulgaria, and Turkey) in the fall of 2010 to conduct interviews with livestock producers about the use of livestock guardian dogs in areas of large carnivore occupancy. We were able to see wolf-fighting breeds in use with both migratory and stationary livestock herds in large carnivore country.

Public support for livestock production in Europe is widespread, and we wondered why. It became evident that it is because European Union governments—and society in general—acknowledge their interest in maintaining livestock grazing and agricultural production. Between 2005 and 2013, Spain lost 28 percent of its domestic sheep flock, while Portugal lost 29 percent, largely due to economic conditions. Government officials fear that without subsidies, meat and livestock production will collapse. Some of the areas in the countries we visited suffer from the "ghost village" syndrome, in which residents have fled rural lifestyles, leaving areas that were once thriving agricultural communities virtually uninhabited. In Spain we saw mountainsides covered with thick heather because livestock grazing and agricultural production had been reduced or eliminated over the years, a condition European Union nations are now attempting to reverse.

The European Union subsidizes livestock production because without subsidies, livestock grazing is not sustainable. Essentially, European society pays to maintain livestock grazing in order to maintain a suitable level of meat for consumers. Livestock production is subsidized at a rate of about $40 USD per head for sheep. The subsidy is higher for producers who graze flocks in the mountain highlands, and even more funding is provided as an incentive to livestock producers to use livestock guardian dogs.

In contrast, in the United States producers pay to graze on public land and there is little public support expressed for this

legal, though highly regulated use of public land, and compensation may or may not be provided for damages inflicted by federally protected predators. If compensation is provided, it is not by the federal government.

In European Union nations, animal production regulations are based on the principle of individual traceability in order to ensure consumer health. For example, government officials in Spain count and microchip all the sheep owned by each producer, and each producer is provided an annual subsidy based on the number of livestock owned.

Tight controls on animal carcasses were implemented in the wake of Europe's Creutzfeldt-Jakob disease crisis in which more than two hundred people died of the disease after consuming contaminated meat. In the past, villages in Spain had pits in place for all farmers from the area to bring their dead animals, and these pits were the main food source for wolves. But since 2002, when the cattle disease Bovine Spongiform Encephalopathy (commonly known as mad cow disease) was detected in Europe, rules were implemented requiring that dead animals be inspected and hauled away by a disposal company, and the carcasses later burned. The government pays 60 percent of the cost of carcass removal, but because the government tracks how many livestock are owned by each producer, producers have to account for all livestock at the end of the year.

The new carcass disposal policy resulted in an 70 to 80 percent decline in the food supply for endangered vultures in Spain. Vultures began attacking lambs and calves at birthing time, and initially biologists denied such activity was taking place. Livestock producers were able to record the incidents with their cell-phone cameras, and researchers soon learned that near-starvation had forced vultures to switch to live prey. Eventually, the European Union relaxed its regulations in some areas so that pits could be created that are only accessible to avian predators. But the disappearance of carcass pits had a negative effect on wolves as well. The closure of livestock carcass dumps eliminated an important food source for wolves, which changed their behavior to not just killing more wild animals, but more cattle as well.

All dogs there must be micro-chipped (€80/$88 USD, paid for by the producer) and have an annual rabies vaccine. With mandatory microchips and rabies vaccinations, local veterinarians host annual "field days" where they drive out to central rural locations to meet producers and see to their veterinary needs.

One evening while we were in Portugal, we shared a very pleasant dinner conversation with a couple from France. Although two of their adult children have become vegetarians, their father proudly referred to himself as an avid carnivore. They explained that in France, beef is produced for the local market. An eartag attached to each calf contains records of its date of birth and source of origin, and remains with the calf its entire life. The calf remains with its mother for nine months, before it is weaned, fattened, and sent to slaughter. When the meat is purchased from the local grocery store, the farm's name is listed on the package of meat. If a consumer prefers that meat, he can target his purchases to that farm's products. The couple from France was very proud of the cattle their local farmer produces. When there are issues (either in terms of resource management or political debates) that would impact the farmer, that producer has political support from the local people who eat his beef.

This is in stark contrast to the situation in America, where the consumer rarely knows the producer whose livelihood depends on the consumption of his product.

Throughout our travels in Europe, we saw shepherds with livestock flocks at all times, but not necessarily to protect the animals from predators. Their purpose is to keep the livestock out of grain or crop fields because there are few fences. When the shepherd is ready for lunch, his two hundred or so sheep are turned into a pen for resting and *siesta* before going back out to graze again later in the day. We also saw herders with burros and cattle, again for the same reason. More cattle were kept in fenced areas, but most of the areas we visited were unfenced, and most herding is conducted on foot, not on horseback. We saw few herding dogs used in any country we visited.

Although agricultural systems in Europe are much less efficient than similar systems in the United States, we found that difference to our liking, as the result is that more people are involved in, and employed by, agriculture. Rather than crop harvests involving large pieces of mechanical equipment, much of the harvest is conducted with smaller equipment and human labor. Many of the livestock production systems we observed were small subsistence operations. Local diets rely heavily on locally produced foods, especially milk and cheese products from sheep and goat production. The average citizen throughout the regions we visited is more familiar with food production and livestock grazing than most Americans. Most citizens had parents or grandparents recently involved in agricultural production. Society overall seemed to be more aware of pastoral lifestyles than the American public.

A shepherd with his flock on a rainy day in Bulgaria.

None of the countries we visited have widespread visible wild ungulate populations for wolves to prey on, as is typical in Wyoming and other areas of the American West. Where there are higher game populations, there are higher densities of wolves.

When Jim and I departed the United States, we initially traveled to Portugal, hoping to learn about the Transmontano Mastiffs we'd read about in that country. We traveled to Montesinho Natural Park, the heart of northern Portugal's wolf country. The park is a native forest of oak and chestnut, with wild boar and roe deer, as well as about thirty Iberian wolves that are part of a larger population roaming in both Portugal and Spain. The park encompasses

dozens of small villages, whose residents are small-scale farmers and herders. Unfortunately, the livestock herds already returned to home pastures for the fall and winter, and we were unable to make contact with enough experts and shepherds to be able to fairly assess the effectiveness of their native livestock protection dogs, although we did encounter a few of the dogs.

Walking the streets and roads of the natural park, we met up with our first guardian dogs, comprising two native breeds. The largest of Portugal's livestock guardian dog breeds is the Transmontano Mastiff, still bred almost exclusively by shepherds for use as working guardians. They have a reputation of being quite reserved and docile, while not being highly aggressive. Work

A Transmontano Mastiff in Portugal.

is being done to gain international recognition for this breed.

The Transmontano Mastiff originated in a pastoral livestock system where stock are grazed in uncultivated areas away from villages, with the continuous presence of wolves leading to the dog's functional massive body structure with long head and limbs, which enable it to travel with the herds, according to the writings of Carla Cruz, a biologist and owner of Aradik Kennel in Portugal.[23] Ninety-five percent of the northern Transmontano livestock protection dog population is still used to protect extensive sheep flocks from wolf predation.

An aggressive program to reduce wolf predation on sheep and cattle herds in Montesinho Natural Park began in 1994, placing Transmontano Mastiff pups with herdsmen. The park maintains a registry of mastiff litters and makes these dogs available to producers. Since the program's inception, the result has been a decrease in wolf depredations on both sheep and cattle.

Government officials pay for wolf damage to livestock in Portugal if there are guardian dogs present, with a guideline of one guardian for every fifty head of livestock, and a maximum of four dogs per herd.

We also encountered a few Estrela Mountain livestock protection dogs. Cruz has written extensively about dogs in her country, noting that in the Estrela Mountains, herds are turned out to graze during the day with only their Estrela Mountain dogs to guard them, returning with the herds in the evenings. Recognizing that most large dog breeds in the guardian dog group have relatively short life spans of about ten years, Cruz writes that local shepherds describe the guardian dog life in this way: "The dog is three years a puppy, three years a dog, and three years a glutton."[24]

Cruz writes that historic records indicate that herders tended to prefer smooth-haired Estrela Mountain dogs, but as wolves disappeared from the mountain in the 1970s and 1980s, this variety essentially disappeared, while the long-haired variety grew more popular with those who desired the dogs as home guardians and companions. The Estrela is probably the most widespread native breed of dog in Portugal. A traditional dog used to guard sheep high in the mountains, because of its beauty, the breed is widespread and often used as pets.

It was interesting to see both dog breeds, and it was notable that we heard concern about the working lineages of these dogs being overtaken by the pet/show lines.

As elsewhere in Europe, the loss of interest in livestock husbandry, combined with a reduction in predator pressure, led to a decrease in livestock protection dog populations in Portugal, but interest has resurged in recent years.

Both Portugal and Spain are two of the five Southern European countries (along with France, Italy, and Croatia) involved in a multi-year European Commission project to improve the coexistence of large carni-vores and agriculture, with wolves and brown bears the major species of concern and conflict. Wolves are responsible for the majority of agricultural damage in all five countries and the bulk of the impacted agricultural producers are those involved in small-scale or subsistence production systems. A central part of the program involved placement of purebred guardian dogs from working parents in areas experiencing conflicts.

Jim and I were stunned to learn of the rather widespread use of illegal poisons in rural areas of Europe. Hunters seeking thriving wild game populations are responsible for placing poison bait stations targeting predators, but the loss of guardian dogs due to encounters with these poisons is substantial. Some agricultural producers also use illegal poisons. Thirty-six percent of the mortalities to livestock guardian dogs involved in the coexistence program in Portugal were due to poisonings.

The European Union Nature Directive provides for strict prohibitions on the killing of certain species, including large carnivores. In recognition of the fact that such restrictions can lead to conflicts with livestock production, European Union guidance documents allow for the use of mitigation measures to reduce conflict, including the voluntary use of guardian dogs where wolves and bears are present.

Throughout our European travels, we noticed a different philosophy about conflicts with carnivores, and the importance of maintaining pastoral lifestyles. Bulgaria provides a good example. The sheep producers and guardian dog advocates there are the same people who are promoting large carnivore conservation. These livestock producers are paid to graze their herds in national parks and natural areas, and they

Estrela Mountain Dogs, because of their beauty, are becoming increasingly desired as companion animals and pets rather than working guardians.

are paid more if they use livestock protection dogs. In most cases, no preference is given to wild carnivores over livestock. In that area, it seems that the public view is that it is just as important to maintain livestock as wildlife.

Throughout Europe, we saw the deliberate grazing of livestock in natural areas and national parks to maintain ecosystem health and benefit wildlife. It is recognized that in order to have a complete ecosystem, all components of the ecosystem must be present. Those components include livestock, the large carnivores that want to eat that livestock, the guardian animals that actively challenge predators that dare to enter the flocks, and the human herders who live amid all these components. This is

Old World Traditions

such a refreshing perspective, and it's one I envy: a societal appreciation of the *morra*.

We also saw parts of Spain and Portugal where livestock grazing has been removed, with a resulting decline in biological diversity. The recent catastrophic wildfire seasons in the West have demonstrated the need to use livestock to help control fuel loads.

I believe that widespread livestock grazing in the presence of large carnivore populations requires the use of livestock protection animals to remain economical. It is my fear that if we fail to enable producers to use guardian animals, and fail to provide systems that allow for continued livestock grazing, then we will have regular catastrophic wildfires due to the build up of fuel loads. The public should be just as committed to maintaining livestock production as it is to maintaining large carnivore populations. If we lose livestock grazing, large carnivore populations will suffer.

Livestock guardians can be *the primary determining factor* between economic success and failure for a livestock producer in large carnivore country. I know that this is true for my own family's operation. There is no way we could continue ranching in the face of the predator challenges we face every day without the work of our livestock guardian dogs.

An Estrela Mountain Dog in Portugal.

The importance of good livestock protection dogs can't be discounted. In Spain, Bulgaria, and Turkey, none of the livestock producers we visited shot at wolves. In some areas guns are illegal, but in Turkey we were told that to kill a wolf with a rifle would be an admission that your guardian dogs simply aren't good enough to do their job.

It's also worth noting that just because livestock producers weren't killing wolves in the areas we visited, that doesn't mean they weren't being killed. Sport hunters were actively killing wolves in many areas we visited. We were told that if you see wolves, they are too bold—they must retain their fear of humans.

The protection of guardian dogs allows for grazing animals to utilize areas inhabited by a variety of predators.

CHAPTER 10

WOE TO THE WOLF

Spanish Mastiffs, male (left) and female (right), on rangelands in central Spain.

Henry S. Randall described guardian dogs as fierce defenders of their herds, writing that "after night-fall the dogs separated themselves from the sheep and formed a cordon of sentries and pickets around them—and woe to the wolf that approached too near the guarded circle! The dogs crouched silently until he was within striking distance, and then sprang forward like arrows from so many bows. Some made straight for the wolf and some took a direction to cut off his retreat to forest or chaparral. When overtaken his shrift was a short one."

Randall's description was penned in 1863 and was in reference to the early livestock protection dogs that were imported with Merino Sheep into America. The dogs were Spanish Mastiffs.

Biologists Yolanda Cortés and Juan Carlos Blanco were Spain's representatives to the European Commission's four-year, five-country LIFE COEX program to reduce conflicts between livestock and large carnivores, and Jim and I were fortunate to spend time with them in Spain. The COEX program not only pays for placement of high-quality Spanish Mastiffs, but also for installation of electric fences and other mitigation and education measures for livestock producers impacted by wolves.

In Spain, compensation for large carnivore damages are linked to the use of damage prevention measures (such as electric fences and mastiff pups). To enroll in the program to receive free mastiff pups, three conditions must be met: the producer must have suffered wolf attacks on livestock; the ranch must be located inside the designated wolf distribution area; and the rancher must be willing to cooperate with the program.

Cortés noted: "Farmers who have not had dogs do not know how to train the dogs. The project not only gave the dogs for free, but also helped teach the farmer to train the dog. I think that's very, very important."

The Iberian wolf population is located in the connected region of northeastern Portugal and northwestern Spain. This portion of Spain has long been home to a wolf population, but the region south of the Duero River was recently recolonized by the predator. Spanish wolves weigh from sixty-five to ninety-nine pounds, with an average

weight of about seventy-seven pounds. A typical wolf pack consists of five to seven wolves, primarily preying on deer, wild boar, and carrion. Spain is home to about two thousand wolves.

Cattle are the most common livestock in northern Spain. "Cattle spend the whole day alone in the mountains, and mastiffs are their only protection there," Cortés explained.

Wolves are recolonizing Central Spain's wide open, rolling agricultural landscape, hiding in the cover of remnant forests, moving down draws and gullies to hunt. Wild ungulates are nearly absent from this region's highly modified agricultural landscapes, and as a result, 75 percent of the wolf's diet biomass consists of livestock carrion.

Most of the grazing areas we visited in Central Spain are unfenced, so the herder must stay with the flock in order to keep the animals from entering grain/cereal fields in the area. The herder stays with the flock until he's ready to eat lunch at about 2 p.m. At that time, the flock is placed in a centrally located pen, which in some cases has been reinforced with electrical wire to keep wolves out. The herder goes away for lunch, and comes back to let the sheep back out after a few hours. The sheep continue to graze, with the herder alongside, until it's nearly dark and they are most often penned again. Larger herds (we saw one with about one thousand head of sheep and eleven mastiffs) are not night-penned, but stay out with their mastiffs. Nearly without exception, the herders are the owners of the animals and the ranch.

As we began our interviews with livestock producers, we quickly learned that Spanish Mastiffs are not called livestock guardian dogs, protection dogs, or simply dogs, but instead are always referred to as mastiffs. To the Spanish livestock producer (sheep, goat, or cattle) there is simply no other animal comparable to the mastiff. It should also be noted that producer preferences are very specific to Spanish Mastiffs, because other breeds sharing the mastiff name are not the same.

Spanish Mastiffs are massive, and most producers allowed us to pet and handle their dogs. The dogs were very tolerant, but quickly went back to work after greeting us. We met dogs that had actively fought wolves, including one female that was still healing up from a battle a few months prior, as well as a big male dog that had killed a wolf.

Mastiffs used for guarding livestock must be from livestock breeders, not just mastiff breeders, we were repeatedly warned, with the added caution that it is unacceptable for the dogs to have human aggression. It was also pointed out on numerous occasions that the presence of a set of double back dewclaws is a characteristic of purebred Spanish Mastiffs and is highly desirable.

Ranchers are compensated for damages caused by wolves in Spain's Castilla-Leon region. In the 1970s, wolves were located north of the Duero River, but as the population increased and recovered in that region, protections were eased. In southern Spain, where there are a few smaller scattered populations of wolves, the animals are still granted protection under both Spanish and European law. When wolves attack livestock south of the river, a team confirms the kill and provides documentation for compensation.

"The wolves and the wildlife here have been shot at for centuries—that's why the wildlife in Europe are much more shy than in America. All the animals are shy," Cortés said. "The wolves are completely nocturnal. The wolves don't howl during the day; they are very shy. In many parts, they don't even howl, not even during the night. They completely avoid humans." Even radio-collared packs with known territories fail to respond to howls from biologists tracking them, a common survey method for wolves in America that simply doesn't work in Spain.

Firearm use is tightly regulated in Spain, and as one producer told us, he is not allowed to shoot at either a wolf or a stray dog attacking his sheep, but instead must call a ranger to shoot the offending animal.

Wolf attacks can have both long-lasting and unexpected impacts on a flock. Cortés reported two cases of sheep flocks that were attacked by wolves, with these herds subsequently becoming afraid of dark-colored mastiff dogs. The COEX program replaced the dark-colored dogs with light-colored or white mastiffs to get the flocks to once again accept the presence of guardian dogs. It is assumed that the dark brown and black dogs reminded the sheep of wolves.

Cortés said program participants in Spain have found that some of their mastiffs have been killed by wolves, as well as found wolves that had been killed by mastiffs. Most of the dead mastiffs were females killed near wolf rendezvous (pup-rearing) sites. Although physical fights between mastiffs and wolves are not frequent, it is also not a rare event. For the most part, the mastiffs are much larger than the wolves, and barking is usually enough to deter wolves from protected flocks. The wolves then turn to unprotected flocks as prey.

When asked if she believed that wolves are attracted to an area by the presence of livestock guardian dogs (perhaps because of competition or the potential for breeding) as some American biologists have sug-

Wolf biologist Yolanda Cortés greets an eleven-month-old Spanish Mastiff pup.

gested, Cortés responded, "I think that's nonsense." Cortés pointed out that it is easier to find a female dog than a female wolf, but the cases of a wolf trying to breed a dog are low. She said wolves may approach dogs either when wolves are at very low densities, or in response to a female dog's heat cycle.

Goya

Goya is a milk sheep producer in the Segovia area of the Castilla-Leon region of central Spain who had frequent incidents of wolves attacking his sheep at night while the sheep were in the field. The situation had become so extreme that Goya was spending nights out in the field with his sheep in order to protect them.

After wolves arrived in the area and attacked his flock, killing twenty to twenty-five sheep, Goya began using mastiffs. "The instinct of the mastiff for protecting livestock is very strong," he said. Although he expressed no preference for male mastiffs or female mastiffs, Goya noted that in his view, males are more territorial, and usually defend territories much more than do females. Goya said the number of dogs recommended is dependent upon the area where the livestock is grazed, pointing out that one thousand sheep in an enclosure is different than one thousand sheep on the mountain.

Cortés said that most ranches include eight hundred to one thousand sheep, many of them with only two or three mastiffs. She noted that there is less threat of predation in Spain than in Northern Rockies with coyotes, so it's logical American producers may need a higher number of dogs, especially in range flocks.

Training is very important, Goya said, recommending producers be very strict with pups during training. He noted the importance of constantly controlling the dog: If the dog goes away, it is important to call him back and teach him to go to the sheep. "The dogs must stay with the sheep. They must always stay with the livestock," Goya said. We saw that Goya's dogs were attentive to the sheep, licking their faces and demonstrating affection.

"The most important thing is that you have the control over the dog," Goya said. "The dog must learn that you control the situation; you must teach them to stay with livestock; and you must teach him that you are the one who rules." Goya recognized the problems for range producers in establishing this type of relationship with the dogs, suggesting it was easier in a farm-flock, or stationary type of situation.

When we approached Goya's farm in the car, a mastiff ran up to the car barking. This dog was associated with a small flock of sheep that was passing by Goya's farm at the same time that we arrived. Goya would not allow his dogs to approach cars. If his dog behaved in this fashion, Goya said he would go to the dog and reprimand it for such behavior. If the dog were allowed to proceed without correction, Goya feared it might become bolder and eventually show aggression toward a human. The dogs from the passing flock and Goya's farm were not overly aggressive in their interaction, although Goya said sometimes more aggression is exhibited. The dogs quickly establish and avoid each other's territories, although Goya added that some farmers encourage their dogs to challenge other dogs they meet in trailing or grazing.

Rufino

Rufino owns about two hundred head of Israeli milk sheep in the Castilla-Leon region, an area that contains 50 to 60 percent of the wolf population in Spain, with a relatively low human density. It's a beautiful countryside, thick with Spanish lavender (which smells like lavender, but looks like sagebrush). In this area, shepherds always accompany sheep flocks because the herds graze near grain fields where there are no fences, a common scenario in Spain. Typically, his sheep graze during the day, are penned at midday, and are freed for afternoon grazing. The sheep are night-penned and are guarded by at least three mastiffs that roam outside the pen, which is reinforced with five-wire electric fence.

Rufino had used livestock protection dogs in the past, but in the last five years has switched to breeding pure Spanish Mastiffs. He uses mastiffs as a prevention meas-

A sheep herd with its livestock guardian dogs passed another sheep farm with its Spanish Mastiffs. The dogs met and mixed momentarily without undue aggression.

ure, and reports he has had no problems with wolves since acquiring the dogs.

Although his dogs are not aggressive toward humans in the field, Rufino reported they but do become more aggressive when located at his stables. Although he uses herding dogs as well, Rufino reported no conflict between his mastiffs and the herding dogs.

Francisco

Francisco is a meat sheep producer who runs 1,500 head of sheep of the native Spanish Castellena breed, accompanied by four mastiffs. He had one wolf attack before acquiring "good" mastiffs, but since using "good" mastiffs, no attacks have occurred. He explained that by "good" mastiffs he meant purebred dogs from other working producers.

"Mastiffs are very, very useful for not just wolves, but stray dogs," Francisco said. In his flock we saw one sexually intact male mastiff with a naturally docked tail that was very shy to us strangers. Two other male mastiffs in his flock had been castrated to prevent them from roaming.

Francisco reported no general problems with dog aggression against humans, but like Rufino, he noted that his dogs are more aggressive toward humans near the stables, and not out in the field.

A two-year-old male Spanish Mastiff watches as a six-month-old mastiff visits with the author.

Carlos

Carlos is the foreman at Caserio De La Torre, a large range operation similar to those found in the beautiful rolling range country of central California. We arrived at the Caserio De La Torre to learn that a pack of wolves lives on the ranch, attempting attacks on livestock herds on a near-constant basis. The introduction of mastiffs resulted in major declines in attacks on livestock, although some do still occur.

The night before our arrival, a wolf had jumped over an eight-foot-tall, wire-topped rock wall and killed an adult sheep in a pen, as we later went out to observe and confirm. The carcass had been left in place, since a ranger planned to use it as bait for shooting the wolf should it return during the night.

Livestock losses to wolves on the ranch have declined over the years, but are still significant:

- 2007 – 27 calves;
- 2008 – 11 calves;
- 2009 – 24 calves; and
- 2010 – 12 calves and 8 sheep;

After suffering substantial losses in 2007, the ranch was first presented with three mastiffs, but the dogs had been bonded to sheep and not to cattle and were not successful. Carlos came to the ranch in early 2009 and brought his own mastiffs with him. As we talked, we learned he was in the process of liquidating the ranch's three hundred sheep (unrelated to wolf issues), and converting entirely to a five-hundred-cow production

Livestock Guardian Dogs

A Spanish Mastiff with a mid-length naturally docked tail on a ranch in Spain.

outfit, with five adult mastiffs and two pups for protection. The ranch breeds Avileña cattle, a native Iberian breed that is known for its bravery and that produces high-quality beef. Carlos showed us photographs of wolves approaching his cattle, only to be confronted and chased off by his mastiffs.

Wolf biologist Yolanda Cortés explained that the mastiff guarding behavior generally is to chase the predator away from protected livestock, but after a few hundred yards, the dog should stop and return to the herd. She said producers should avoid chase behavior in their mastiffs.

Carlos led us on a tour of the ranch's bonding pen containing yearling cattle and two young mastiff pups. The pups were allowed constant access to the cattle in a building, but could escape into their own pen for protection if needed.

Cortés explained: "You must put the pups together with cows, in a close place for four to six weeks more or less to create the social bond. It's more important to create the bond with cattle than with sheep. You must be more careful with cattle, but the process is the same."

Carlos began using *carlanca* collars (spiked collars) after wolves tried to kill his mastiffs. Four months prior to our visit, the shepherd arrived to find the adult pair of mastiffs "had been beaten by wolves." It was nearly two weeks later before they found the dead wolf in the pasture, proof that the dogs hadn't lost the battle. Wolf attacks occur year-round on the ranch, but late summer and fall are most common, when wolf pups are getting bigger and starting to hunt more with their parents.

The author with a one-and-a-half-year-old male Spanish Mastiff and his owner.

The ideal number of dogs with free-ranging cattle is dependent upon the terrain in which the herd grazes in addition to the size of the herd, according to Carlos. He prefers to have at least four dogs with three hundred head of sheep, and added that the cattle are always available for wolves, since they are continuously out in pastures.

Carlos reported no problems ("no, never") with aggression toward people, even with the presence of a public road and recreation area within the grazing range: "Mastiffs are selected to confront predators, not people." He reported that leashed dogs accompanying recreationalists are not a problem, but unleashed dogs are, since the mastiffs will confront them. Hunting dogs are also a problem, as wild boars are hunted in drives, and the hunting dogs are sometimes lost on the range. He explained that mastiffs kill dogs that attempt to come into the herd—this was the fate of two hunting dogs that entered his herd.

A bonded pair of Spanish Mastiffs on a large ranch in Spain. The male (right) had killed a wolf in defending the ranch's herds.

Cortés cautioned that mastiffs that show untoward aggression should be trained to reinforce that such behavior is unacceptable. Carlos recommended producers focus on bonding their pups, begin their education early, and correct undesirable behavior from the beginning.

Guadalajara

Wolf biologist Juan Carlos Blanco guided us to livestock producers in the Guadalajara area east of Madrid, in the Castilla-La Mancha region. There are a few brown bears in northern Spain, but not in Central Spain, so the major predator of interest there is the wolf.

"Wolves are new to this area—breeding packs were first documented here five years ago," he said, so many producers are now learning how to use these dogs.

"Some producers don't like *carlancas* because the dogs fight. They can be injured by the *carlancas*," Blanco said. He noted that female mastiffs are very territorial and competitive. While some view competition among the dogs as a problem, he said, "Some say it's nice too," since the female dogs will spread themselves out around a herd to avoid each other.

The average number of wolves in the local wolf pack he studied in 2010 was ten wolves. Blanco said that a research program radio-collared seventy wolves, and there were ten known deaths of those wolves. None were killed by other wolves, but a mastiff in a cattle pasture killed one wolf, and the remainder were human kills (both legal hunting and illegal killing).

Wolves killed a mastiff on a large ranch where two thousand sheep and three hundred cows are kept along with their twelve adult mastiffs. The sheep spend nights alone in the field with their mastiffs, and wolves are around all the time. Blanco said that good mastiffs stay with sheep and the biggest management problem posed by the dogs is when a female dog enters her heat cycle.

The livestock herder must be sure all sheep are together, or are in two groups with mastiffs in both herds, before he goes home for the night. Wolves kill some sheep from time to time, but not many, he said.

October/November through January/February is the wolf-hunting season for the area, with a 140-wolf quota in the Castilla y León region, according to Blanco. In wolf damage areas, additional wolf control is authorized. There are 149 known wolf packs in the region, and wolf hunting is concentrated in wolf damage areas. Cars kill many wolves, but at least half of wolf mortality is due to illegal hunting (according to a ten-year-old survey), Blanco noted.

In northern Spain, wolves never disappeared, and neither did the mastiffs used to guard against them. Because of this long tradition, the mastiffs protecting the herds became simply part of the landscape. In the Basque country, wolves arrived within the last twenty years, and in response, dogs were reintroduced. Trouble with tourists resulted, he said. People in agricultural areas of Spain know that they have to be careful around livestock, so most members of society are more familiar and accepting of mastiff dogs.

In Spain, there is now a working branch and a show branch of mastiffs, Blanco said, again emphasizing that it is important that livestock producers only use dogs from the working lineages.

The importance of using mastiffs was underscored by recent livestock depredations in the region. As we drove near el Pico del Lobo Mountain, we learned that the day prior, wolves had killed six sheep on the other side of the mountain. A month earlier, wolves killed sixty sheep of a herd of 150 that had been left unprotected, Blanco said.

Juan

Juan Arenar is an influential livestock producer from the village Cantalojas. His operation consists of two thousand sheep and three hundred cows, accompanied by a local pack of wolves that keeps Juan's sheep flock surrounded. His sheep graze on rangelands governed by communal/ancestral grazing rights, administered by local towns. The sheep are herded during the day, or checked by a herder a few times a day. The dogs are fed in the mornings, and sheep are checked late in the day to be sure they are guarded and in a good area for bedding. The flock grazes a "natural area," or "natural park." The Tejera Negra Beech Forest Park was designated a Natural Site of National Interest in the 1970s to protect the beech forest in the southern portion of the Iberian Peninsula.

Juan's dogs come to the approach of the truck every day for food. It's a typical sheepherder truck, dented and dirty, a bag of meat scraps and bones in the back.

"They protect the sheep in exchange for food," Blanco explained of the relationship between the herder and his mastiffs. When a sheep dies, the dogs protect the carcass, but there are many vultures in the area, and they work very fast, Blanco said.

Juan said he has used the mastiffs for more than ten years, and trades other live-

Inside Juan Arenar's barn, where sheep are sorted into lambing pens and mastiffs are free to visit and mix with each group.

stock producers for dogs when needed. He believes the dogs have improved his sheep's behavior, making them "more compact" or better at flocking.

When the mastiff pups are a few months old, they are placed with sheep, where they remain for an intensive bonding process. Some pups will try to play with the sheep and will wound them. When this happens, Juan Arenar advises that the pups be placed with male sheep (bucks/rams) for a month so they will learn to behave, since the rams will not tolerate such behavior. He noted chase behavior often happens with yearling pups, which must also be corrected. But pups that are gentle with sheep can go out onto the range with sheep when they are six months old.

Just one dog on Juan's range was wearing a *carlanca*, although he told us he prefers for his dogs to wear the collars. We saw several of his mastiffs that were limping because they had been fighting over a female in heat. When I questioned why one female dog was wearing a bell, I learned that the dog was pregnant. She would have pups out in the grazing area, and the bell would allow the herder to track her and check her pups. When a litter of pups is born, there are usually five to eight pups, but only three or four at most are saved. Juan said he always selects for double back dewclaws because this is a sign of pure mastiff.

The only health problem identified with Spanish Mastiffs is hip dysplasia, although it was reported that dogs with this condition are from heavily inbred lines, not working dog lineages. Another occasional health issue reported is that the stubble on cereal fields will rub against their back dewclaws, causing chafing. One of the main problems with Juan's dogs is that hunters kill them, a problem we heard repeatedly in both Spain and Bulgaria.

Juan reported that some of the dogs will run after wolves that are detected during the day, while others will not. At night, he reported, the dogs stay closely alongside the flock. When we visited Juan's barn, we saw a female mastiff that had been attacked by wolves in late summer. She had been very sick afterwards, and she was still recovering two months later. Any sheep that wander off on their own, or in small groups, will immediately be killed by wolves, Juan reported, adding that on one occasion, a group of twenty sheep was released without mastiffs and was promptly killed.

Paulino

Paulino is a goat producer who uses six mastiffs with his main herd of 448 goats. His herd primarily gives birth in his barn, but sometimes out in the fields. Paulino's herds were grazing on acorns when we visited.

Paulino's first ordeal with wolves was only three years prior, when a wolf pack moved into the area. There were at least two wolves involved, but the exact number is not known. Many livestock producers didn't believe there were wolves in the area, and few precautions had been taken.

Paulino had two mastiffs at that time, including a very pregnant female, and a ten-month old pup, so his goats were not well guarded. The wolves killed fifty-one goats in one night. After that depredation incident, Paulino purchased several more mastiff dogs and has not suffered any major losses since then.

Paulino often locks his goats in the barn or a pen at night, but some nights the goats stay out with their mastiffs. He factually noted that any goat left outside without a mastiff would be killed by wolves. "The goats must be protected," he said.

Our last visit one afternoon was to Paulino's second herd, consisting of sixty mother goats that were out grazing away from their penned kids. Paulino reported there was one mastiff dog with this bunch that he would like for us to see. We arrived at the pen in the evening, and the mother goats were nowhere to be found. We walked through thick brush covering the mountainside, trying to find the herd, but couldn't even hear their bells. Paulino decided to drop down into the canyon below in attempt to find the herd and place the mothers back with their kids for the night in the safety of the eight-foot tall wire pen, so we were to wait near the pen until he returned.

As it started to get dark, and we could hear the goat bells ringing in the distance as the herd moved closer, we (wolf researcher Juan Carlos Blanco, Jim, and I) walked back to the kid pen, opened the gates to let the mother goats in, and stepped back out the way. We realized that if the goats tried to approach the pen and saw strange figures in the darkness, they would never enter the pen. So Jim and I stood very still next to Paulino's vehicle, while Juan Carlos stood on the other side. The goats began coming to the pen, but they approached from both

Spanish Mastiff female (left) and male (right) guarding a goat herd in Spain.

sides, so Jim sat down on the ground so he couldn't be seen. Afraid to move, I stood frozen in place.

Suddenly a large male mastiff approached the pen with the front of the herd, as the goats began to enter. The male stuck his nose to the ground and wheeled around looking in my direction. I warned Jim so he could get up off the ground, and began softly telling the massive dog what a "good puppy" he was. The dog barked loudly at me and came directly toward me, but when he approached, he simply sniffed my hands, which I quickly used to pet and praise him. He gently brushed my hands with his teeth, and then passed behind the vehicle to meet

Juan Carlos. I could hear Juan Carlos talk to the dog before the dog continued his circle to meet Jim. The dog softly brushed Jim's hands with his teeth as well, but did not bite.

That was a miracle. We had created a worst case scenario in which we fully expected to be attacked by a guardian dog, yet the dog did not bite anyone, and only showed mild aggression. The dog was very nervous, and although Paulino was talking to us, as we approached the goat pen, the dog continued to brush our hands with his teeth, taking our hands into his mouth in attempt to redirect our attention from the goats to him. Understanding his body language and what he was attempting, we walked away from the pen. This increased the dog's comfort level and he went inside the pen to his goat herd, with we strangers safely locked out.

It was too dark to get a photo, but this was a typical massive mastiff, just over a year old. Paulino's mastiffs were not friendly like others we had met, and did not want to be touched by strangers. This is probably a reflection of Paulino's belief that the dogs should not be petted while they are being bonded to livestock as pups. His largest and most valuable mastiff, Leon, was always nearby, but lurked in the brush where we could never fully view him. Leon was the only dog wearing a spiked leather collar as a defense against wolves. We found that the collars are often reserved for the best dogs.

Paulino uses the same bonding process as most producers in America, placing the pups in a pen of kid goats to bond for their first two or three months of life. Little training is undertaken, although if a pup chews on the ears of a kid, he spanks the pup with a newspaper.

"They should always be with the goats, and with the other dogs," he said. "Even when the pups are young, if they are with older dogs, this is enough to train them." Paulino breeds and raises pups, giving excess pups to other producers. Paulino said he does not castrate his dogs.

"The goats are the family of the dogs," Paulino said through his interpreter. "They believe they are." Although the goats and mastiffs share expressions of affection, the mother goats will also attack the dogs during kidding if they feel their kids are being threatened. One of Paulino's dogs is very protective of any lame or sick goats that tend to follow behind the herd, and this dog will push the slow goats with its head to hurry them along.

Paulino does not pet or play with the pups, and if they approach him for affection, he rejects them. Paulino advises that producers avoid strong links with the dogs, fearing that the dogs will leave the goats and attempt to follow the herder. He does pet his dogs enough so that he can catch them for veterinary care.

Paulino feeds the mastiffs once or twice a day, using kibble and bread mixed with fat. Because the goats will compete with the dogs for food, he supervises the feeding to reduce conflict. Like his goats, Paulino's dogs were a mixture of colors, but Paulino admitted to having a preference for white dogs, as he can see them better from a distance.

Paulino is very happy with his mastiffs, and recommends that American producers in wolf country try the breed. Although he leaves his goats alone on the mountain with the dogs at times, Paulino fears the wolf pack will return.

When asked about mastiff aggression toward humans, Paulino said he has seen his dogs bark at people in a threatening manner, but never bite anyone. Paulino also reported that on occasion his mastiffs have conflicts with his herding dog, but attributed that to the smaller dog's insistence that it is the top dog.

It is worth noting that the livestock producers we interviewed throughout Spain spend much time with their herds, often on foot. Their animals are accustomed to human supervision, especially those herds that graze the lowlands in unfenced areas adjacent to cropland, and that are penned during the herder's midday meal. Herds that graze in the uplands do so in areas of thick cover of trees and brush where hiding cover for predators is plentiful. Farmers in the lowlands see their dogs every day, but only every few days when their flocks are grazing in the mountains. Cortés said, "That's why it is so important to avoid aggressive behavior," since mountainous regions can have high recreational use.

We were extremely impressed with the working Spanish Mastiffs we met in Spain, and came back recommending that livestock producers in wolf country in the United States try this breed. Their effectiveness against large carnivores is highly desirable.

A Spanish Mastiff moves with its herd across the highlands of Spain.

CHAPTER 11

BEAR BRAWLERS OF THE BALKANS

A three-year old male Karakachan at work in the southern Rhodope Mountains of Bulgaria. The dog is wearing the required *spavachka*, or dangle-stick, attached to his collar.

Survival of the guardian depends on the survival of the predator and vice versa." That's the philosophy of Sider and Atila Sedefchev, as they speak about the Karakachan livestock guardian dog and the wild predators the dogs must defend against across the rugged Bulgarian landscape. Their interest begins with conservation of the dog, but extends to a variety of predators because these animals developed and evolved together, and they need each other in order to reach their full potential—to survive with their evolutionary potential intact.

As we considered the possibility of Karakachan dogs, I scanned my memory for anything I recalled about their Bulgarian homeland and quickly realized I knew very little. I considered it a region long plagued by war and struggle under Soviet oppression, which was once welcomed to restrain the Ottoman Empire. The country is a new member of the European Union (since 2007), and the poorest nation in that venture, with an economic underpinning of organized crime. In our travels across Bulgaria, the exodus of people from mountain villages and rural areas was evident, as people gave up rural life and moved into cities. This has occurred for decades, and the resulting afforestation is stunning. Coniferous forests have overtaken parks, clearings, meadows, fields, and homesteads.

The Karakachan people of present-day Bulgaria were nomadic herders, with each family owning from five hundred to four thousand sheep, a herd of horses, and a small pack of guardian dogs. The closure of borders between Balkan countries after World War I resulted in the restriction of movements of these people and their flocks. Nationalization of lands followed in the following decades, with nomads forced to settle and their flocks taken to collective farms under the direction of the Bulgarian Communist Party. Dog eradication was undertaken, and shepherds were allowed a maximum of two dogs per flock. When the collective farms were discontinued in the early 1990s, livestock were slaughtered and many more livestock guardian dogs were lost as they were set free to starve or be killed. During this time, the nation began

the process of privatization of agriculture. It was then that efforts kicked in gear to preserve the native Karakachan guardian dog, with key conservationists realizing that to preserve this dog, all components of its environment should be preserved as well, including native livestock, predators, and the environment in which they all evolved. Karakachan advocates believe that to preserve this guardian breed, the dogs must perform their traditional work in their natural environment.

The Karakachan livestock protection dog is an aboriginal breed of Bulgaria, but official breed recognition under Bulgarian law did not occur until 2005. A program was begun in Bulgaria in 1996 to conserve the Karakachan livestock protection dog and its original type and working abilities, focused on "conservation of predators, livestock, pastures and pastoral traditions: conservation of the unique symbiosis between all these elements." The program places Karakachans with shepherds to protect their flocks from wild predators.

We were fortunate to meet up with Atila and Sider Sedefchev for an interview about large carnivores and Karakachan dogs before we began our journey into the Bulgarian countryside. Atila had agreed to accompany us as our guide, and we would conclude our tour at his brother Sider's farm near Kresna. The Sedefchevs are noted experts on the native Karakachan—the guardian dog aptly called "the bear fighter of the Balkans."

Sider reports that the Karakachan works well in protecting sheep, goats, and cattle from attack by wolves, bears, and golden jackals. Bulgaria has some of the highest large carnivore densities of any country in Europe, with about 2,300 wolves and six hundred bears. A European brown bear

Sider Sedefchev with a litter of Karakachans at his farm near Vlahi, Bulgaria.

Livestock Guardian Dogs

A young Karakachan at work in Bulgaria. The dogs are known as the "bear fighters of the Balkans."

had attacked a woman out picking mushrooms just a few months prior to our visit.

The program to distribute Karakachan pups to livestock producers in regions of the Balkans that are inhabited by large carnivores resulted in an 80 percent decrease in livestock depredations. When Karakachan dogs were placed with certain flocks in an area suffering depredations by wolves, other neighboring herds that remained unprotected experienced an increase in depredations. The wolves had turned to easier prey.

Much of the livestock grazing in southern Bulgaria involves common flocks, in which the livestock is gathered together from all the people in a village and are grazed together. The size of each flock varies, but one flock included 1,200 sheep gathered from 114 owners, and guarded by four Karakachan dogs. The contract for participation in the program requires the producer to give Karakachan pups from future litters to other producers—a sort of "paying it forward."

Herding of animals is mostly done on foot. In much of southern and central Bulgaria, the landscape is generally too steep and rugged for horseback-based herding. The close association between the shepherd and his flock is evidenced in a tradition explained by Atila. The shepherd cuts the thick wool of a sheep in his flock to use as a pocket for holding the shepherd's money and small valuables. Should a robber seek to mug the shepherd, he would not succeed in stealing the valuables, since only the shepherd knows which of the sheep in the flock possesses the wealth.

We found Karakachan dogs to be extremely athletic, and learned that they tend to work together as a pack to move danger away from the herd. The bear population is dense in this region, and the dogs do an effective job keeping the bears out of the herd. Physical fights between Karakachans and carnivores in Bulgaria are not rare. Sider said that when a wolf approaches, the Karakachans give chase. "We have seen that if a wolf stops for a moment to scare them, the dogs go in directly, fighting with a clear intention to kill the wolf. However, it is very unusual that dogs manage to catch or kill a wolf. Usually the wolves outrun them."

Aggression varied in the dogs we met. Most Karakachans are used in high-use recreation areas, yet have little conflict with humans.

"The Karakachan dog is strictly territorial," Sider explained. "It accepts the flock as its territory, wherever it is." When a stranger approaches, the dogs become visibly aggressive as a warning not to approach closer, yet when a flock passes through a village, the dogs walk calmly without paying attention to people. But there is another reason for the lack of conflicts with recreationists that encounter guarded flocks in the mountains: The tradition of guarding livestock with big, aggressive dogs has always existed. The public is long familiar with this tradition, and sidesteps the flocks to avoid conflicts. Sider believes his dogs are able to look at people and judge a person's intentions. His dogs react differently

Two Karakachans guarding their herd in the fog. The white-colored dog is six months old.

to new volunteers at the farm than they do to strangers coming from the village. Sider admitted that his dogs have bitten people—people who tried to cross through his grazing flock. People must be taught to go around the herds, he emphasized, because the dogs must be allowed to do their jobs.

While most of the Karakachans we met did not display much aggression toward people, we did meet one breeder specializing in human-aggressive livestock protection dogs because of livestock theft by humans in a nearby settlement.

Although aggression isn't displayed toward humans, these dogs must be aggressive to predators to do their jobs. Atila maintains that the dogs must like challenge to be effective guardians in large carnivore country. Sider agreed: "The dog is brave and has instinct, or not."

As Bulgaria is a member of the European Union, livestock producers receive a subsidy per head of livestock, plus per hectare, at a stocking rate of six sheep per hectare. If the producer uses guardians, the subsidy is higher. Producers are paid to graze their herds in national forests, national parks, and municipal lands. We also learned that taxes for goats are ten times higher than grazing for sheep. Goats are viewed as bad, or "evil," and are forbidden from mountain grazing. Every dog is microchipped in Bulgaria in compliance with European Union requirements.

The thick, brushy mountainsides we observed were once open country (within the last one hundred years), because shepherds would set fire to the countryside when they left for the season. The Bulgarian sheep inventory has been reduced from a high of ten million sheep to 1.5 million sheep today. In our interviews, one shepherd commented that he felt he could not graze livestock without his guardian dogs. He noted that wolves are the same color as the trees and are too difficult to see, so he depends upon the dogs to keep his flock safe.

The most serious problem for guardian dogs in Bulgaria is their killing by hunters, especially through the use of illegal poison baits set out for predators, in addition to direct shooting because guardians kill hunting dogs that try to penetrate a guarded flock. The problem has become so severe that shepherds carry a poison antidote in packs on their backs.

Bulgaria is one of several European countries with a law that requires guardian dogs to wear a foot-long dangle stick attached to their collars. This stick, called a *spavachka* in Bulgarian, is to hang to the elbow joint of the dog. Sider explained the purpose of the *spavachka* is to act as a hindrance to prevent the dog from running after game animals, and a dog without one can legally be shot. Shepherds despise the law and feel it hinders the dog's ability to work. They have devised numerous *spavachkas* that comply with law, but do little to hinder the dog's movement.

Much like the Spaniards, the Bulgaria tradition is to raise only two pups per litter, but since there is an effort being made to distribute Karakachans and more dogs are needed, the culling program has been softened. Traditional cropping of one ear is performed to assist in hearing for working guardians in this region, and is practiced by many, although in some European countries it is illegal to crop dog ears, or dock tails, or to remove dewclaws.

Back dewclaws seem to be a particular fascination with guardian dog owners. In Bulgaria, back dewclaws are known as "wolf fingers" because of the gripping power of the dewclaws (even though wolves don't

The author visits with a young Karakachan in Bulgaria.

have back dewclaws). In Turkey, guardian dogs with back dewclaws are called "wolf killers." In Spain, back dewclaws are a sign of a purebred Spanish Mastiff—the dog most suitable for defending against wolves.

Daniela

Daniela Chakarova of Progled, Bulgaria, has used Karakachans for more than twenty years. Daniela's days are spent at her ski patrol job at Pamporovo Ski Resort in the winter, and trekking those steep mountains in the other seasons as she tends her sheep and goats. In the winter, her herd is located on the mountainsides outside the ski resort village of Chepelare and near Progled in the Rhodope Mountains. The herd grazes in the summer in Hadzhiitza and Karamanitza mountain pastures. She uses seven dogs with three hundred sheep and goats, with three or four extra dogs available. She does not fence her flock, but does use night pens. Sheep are lambed out on the mountain.

Daniela said Karakachans "are very clever." The dogs avoid direct conflicts with bears, instead making the bears run away from the herd. The dogs are very athletic, and move very quickly. Most problems are with brown bears that come into the villages, Daniela said, while most wolf attacks involve more than one wolf. "That's why we prefer to have young dogs," she said, pointing out that old dogs go out after the wolves, and the young dogs stay back and

Daniela Chakarova visits with a red-colored Karakachan near her herd.

bark. Daniela uses iron spiked collars on her dogs, and said she's still using them mostly because it's a Bulgarian tradition.

Daniela said she would not be able to continue in the sheep and goat business without her guardian dogs. Some dogs patrol within the herd, others to the outside, and she knows that if a dog is missing, that means she is missing sheep. Her dogs give birth in the wild, with the livestock. Only the dog's owners are allowed to touch newborn pups, while feeding the female, so they will have the ability to catch and handle the pups later on. The pups become bonded to the flocks and to the shepherd only. The dogs tend to keep their distance from people, but may attack people if they come too close. If you stand quietly and do not enter the herd, you'll be fine, she notes. While we were on the mountain amid Daniela's herd, a man from a nearby village came near the herd while picking berries, and the dogs did not behave aggressively toward the man.

At one year old, the pups will defend the herd, but will have little experience, Daniela said. After they are three years old, the dogs are perfect guardians. She said the dogs live to fifteen years (a few cases of twenty or twenty-one years old). Like most Bulgarians we met, Daniela teaches certain commands to the dogs, including commands to "go to the sheep" and "stop attack."

Daniela has preferences for certain characteristics in her Karakachans. She prefers her dogs to have double back dewclaws, black mouths, and large-sized pups. She

Bear Brawlers of the Balkans

Chakarova's red Karakachan, with its naturally stubbed tail.

advises the use of white dogs in a dark flock so they can be seen as they move in the fog. She likes a contrast in colorations, and doesn't like dogs that are completely white or completely black.

The Varshilovs

We met up with the father and son team of Dimitar and Georgi Varshilov who own a flock of sheep we encountered as it grazed on thistles in a field in a populated area near Plovdiv. The Varshilovs manage their eight hundred dairy sheep in four flocks, with three or four dogs used in each group in the summer on the highlands. Their flocks are sometimes night penned, and in bad weather are kept inside, but in good weather, the flocks stay out with the dogs. Some dogs

stay inside the flock, while others patrol outside, and the dogs load into livestock-hauling trucks alongside their sheep when it's time to move. While the dogs are in the lowlands during the winter months they are chained or tethered. We visited two of the family's four flocks, and enjoyed learning from these livestock producers with four decades of experience with Karakachans.

The primary predators of concern to their flocks are bears and wolves, with bears posing as the lesser threat. Bears don't care about sheep because they are fed at bear feeding stations, they said, and with big dogs, there are no bears. "The dogs know the bears," Dimitar said.

Occasionally there are problems with wolves, such as one year while there was

A one-and-a-half-year old Karakachan near Plovdiv.

Bear Brawlers of the Balkans

only one young dog protecting a flock and a wolf grabbed a sheep around the neck but did not manage to kill it. The family has had no losses on high pastures in the last six years, even though wolves are present.

The Varshilovs noted that Karakachans mark their territory, and believe that the more dogs marking the territory, the better. The wolves learn and do not challenge the dogs. "So the wolves, when they see the dogs are serious, they go on," Dimitar said.

The Varshilovs have not had dogs killed or injured by wolves. They breed their own dogs, keeping five or six pups of an exceptional litter. The dogs bond with the sheep, lick the lambs, and show affection. The dogs will even lick and clean injuries on their sheep as well. Pups move with the herds from a very young age, and pups that are born on the mountain with

the flock learn from older dogs. The shepherds believe that pups born out with the sheep, and traveling with the movements of the flock, lack health issues like hip dysplasia.

This family has a color preference, preferring black and white dogs, believing them to be more visible in the flocks. Their dogs fight occasionally, most often over females in heat. The family has used spiked collars in the past. Although this family does not dock their dog's ears because the men personally do not like it, they cautioned that if the dogs are fighting often (either with each other or with wolves) the ears are the weak spot. They do not castrate dogs either, with the exception of a dog that wasn't staying with the flock. They reported the maximum size of their adult male dogs is about one hundred pounds.

An adult male Karakachan in the mountains above Plovdiv, Bulgaria.

Miroslav

Miroslav Marinov is of Zmeyovo village, near Stara Zagora, Bulgaria. He owns the Volcan Karakachan kennel, and raises 450 head of Zakar sheep, the local sheep breed of southeastern Bulgaria. His herd grazes in winter near Zmeyevo in Sredna Gora Mountain, and in summer, in Central Balkan National Park.

A veterinarian by trade, Miroslav has been successful as a professional musician as well. As his wealth increased, he purchased a farm and livestock fifteen years ago specifically so he could have Karakachan dogs. First, he bred goats on his farm, but later switched to sheep. Because of past problems with human thievery, Miroslav selects for human-aggressive guardian dogs. These dogs are aggressive enough that the dogs will prohibit new shepherds from entering the farm buildings where the sheep spend the night in the lowlands. Five or six "softer" dogs are allowed out with the herd during the day, but the "harder" dogs are unleashed at night. His dogs sometimes fight, but in principle have a strong pack hierarchy. Sometimes the dogs fight for dominance, or for food, and females. The dogs will keep all intruders out of the herd, but must show submission to the flock. The dogs launch very fast attacks, which he said is important when facing bears.

Training a Karakachan to be human-aggressive. The dogs are threatened by a human with a long stick, but are not beaten.

Miroslav said wolves and jackals are his major predators, with bears a lesser threat, but the damage varies from lowlands to highlands. His dogs have deterred six bear attacks that he knows about, Miroslav said. One seven-year old female Karakachan we saw had her leg broken in a fight with a bear three months prior. He finds tufts of bear fur in the mornings, proof of the dogs' encounters with marauding bears.

Miroslav reported that in the highlands, one flock guarded by eight dogs had twenty-five sheep killed by wolves. Another herd had fourteen sheep killed by wolves. The herder used lights and a pistol to make noise to keep the wolves away. He eventually gave up this pasture and got better guardian dogs. One night Miroslav left two dogs with a herd, and eight wolves attacked. Some of the wolves lured the dogs from the herd, while others moved in to attack. The flock separated into three groups, and some were attacked on necks, others on the shoulders. This mountain pasture is within the Central Balkans National Park.

Castrated dogs have only about 70 percent of their working ability, Miroslav said, but added that it was better to have 70 percent of a dog than zero. "It's rare to castrate a dog," he said, adding that castration causes the problem of more fighting between the sexes (males and females fighting each other). "It is better to have fertile dogs, to protect your lineage," he said.

A pair of Karakachans with a burro in Bulgaria.

He does not dock tails, since he prefers the visibility of curled tails in the flock. His partiality is for lighter-colored dogs, with plenty of white coloration so he can see the dogs at night. He said his sheep, which have encountered wolves, do not like dark-colored dogs. Miroslav said the best guardian dog is a female, and he also selects for double back dewclaws.

When it comes to chasing predators, Miroslav said some dogs chase long distances (as far as a half-mile), while others are short-distanced. In principle, the older dogs with more experience stay closer to the flock, and the young dogs chase farther. This behavior is in contrast to that reported in Daniela's dog pack. Miroslav said the more dogs guarding the herd, the farther the chase will go, and fewer dogs stay closer. The dogs must be brave enough to attack a bear, and not just bay at it like a hunting dog, he pointed out.

The size of the dog isn't as important, Miroslav said, since the group works together as a pack when danger approaches. He selects certain dog behaviors for each of his flocks—some dogs will stay in front, others in back; some outside the herd, and some inside. He said this is important because when sheep are trailing, wolves will try to kill from behind. He selects dogs that work as a team.

In every case, when he removes dogs for any reason, there is a problem with predators. The peak of wolf attacks on his flocks occurs in mid-August and in the autumn when wolf pups are learning to hunt. In the national park where his animals graze, tourists are deemed more important than grazing, and there are conflicts.

Miroslav keeps more dogs than necessary to his operation to avoid a crisis situation if dogs are killed or die. "It is better to feed dogs than wolves," he said.

Sider

Sider's Karakachan sheep graze in Pirin National Park in the Pirin Mountains, and in winter, around Vlahi village. Sider pens his flock of 450 sheep at night in an enclosure constructed of brush. An enclosure is also available on the mountain, but if the pen is muddy, the sheep are left out at night. The flock lambs outside in January and February, with the dogs cleaning up the afterbirth. If a sheep dies out on the mountain, the wolves will clean up the carrion. The dogs will eat the dead sheep if the shepherd cuts it up for them.

Most Karakachans die from being poisoned or shot, but otherwise can have rather long lives. Sider has had two dogs live fifteen years, and one lived to be eighteen or nineteen, but the average is six or seven years for these working dogs. He has had thirteen working Karakachan dogs killed by hunters in ten years. The situation is dire enough that his shepherds and volunteers carry an antidote to poison in their backpacks so they are prepared to provide care to poisoned dogs.

Puppies can be effective guardians, even as young as four to six months old, Sider said. "They can be effective too, because they can see, they can smell, they can give signals to the other dogs," he said. But a dog aged one and a half years "is very effective . . . I think the best dog is a three- to five-year-old dog," Sider said, calling them prime guardians at that point.

When asked how far the dogs will pursue when chasing wolves, Sider said it varies, but sometimes the dogs will run twenty minutes non-stop through the forest, going farther if there are other dogs remaining with the flock. Sider also noted that Karakachans are rarely found in the middle of their flock, instead preferring to surround the herd.

Sider Sedefchev with Karakachan dogs at his farm.

Livestock Guardian Dogs

One of the Sedefchev dogs at work in a mountain meadow.

Sider maintains there are important characteristics of male and female Karakachans, including that the females bond more closely, but are not as good at attacking as males. He maintains that the presence of females improves the males' performance, but having more male guardian dogs is better than having more females because there will be less fighting. Sider said bonded pairs stay together, and older pups help care for younger pups. He said two females with four males is a good combination.

Sider does not castrate any of his dogs, saying that castrated dogs become lazy. "Old shepherds really like castrated dogs because they are not problematic," he said, because the males do not fight and pursue females. He brings the females to the farm for breeding because the males will fight over the females on the mountain.

Sider does not cull pups from litters. Although he doesn't select for color, his preference is for black dogs with white neck collars, and he likes the dogs to show a third color as well. "In the forest, the spotted dog is more visible, during the day or during the night," he explained.

Most of the pups are born at the farm, but if born out in mountain pastures, he follows the female to find the pups. He tries to limit the amount of contact the pups have with humans, but this does not apply to the shepherds. There are many people around the dogs, including volunteers, people coming to visit and see the large carnivore

Bear Brawlers of the Balkans

A litter of Karakachan pups in Bulgaria eating from a communal bowl. Several of these pups had naturally stubbed tails, one had part of its tail chewed off by its mother at birth, and several others had long tails.

center, tourists in the mountains. Limiting the pup's contact with others is aimed at keeping them from following visitors and bonding with strangers. Sider touches them and bonds with the pups, but doesn't want visitors touching the pups. When the pups

Large Carnivore Conservation Center

Sider Sedefchev's farm is home to a variety of livestock breeds native to Bulgaria, but the farm also has a larger role. Sider, along with brother Atila and wife Elena Tsingarska, founded and operate the Large Carnivore Conservation Center on the farm. The center is an educational facility complete with two captive wolves and a European brown bear, in addition to the native livestock and guardian dogs. Elena is Bulgaria's primary wolf researcher, a job she's held since 1997. All three members of this family are committed to large carnivore recovery in the Balkans, and the center showcases the use of Karakachan livestock protection dogs in order for all species to co-exist.

do something bad, he disciplines them, and he doesn't want the pups to go with the nice stranger instead of him.

Wolves are the primary predator that most of Sider's flocks encounter. Most encounters involve two or three wolves, often from a larger pack in the area. Sider noted that when it comes to Karakachans, size really doesn't matter because the dogs are very easy moving and quick in their actions.

"Once the wolves killed a young goat in front of me. I saw the wolf for a second, but it was too late," Sider said. "That same year, our dogs managed to kill a wolf. Some year, maybe four years ago, they killed another wolf, mainly when the wolves were alone."

He was losing fifteen or more sheep a year to wolves in the past. One year while a female dog in heat distracted the dogs, wolves killed three of his sheep. In one year, wolves killed seven young horses and three cows in ten days. The male wolf was eventually killed. The horses and cows were free ranging, without guardians.

"With the sheep, almost nothing in these ten years," Sider said, "because the sheep stay always together. And I have always better dogs with the sheep." The reason his losses are so low is because the sheep flock well when paired with good guardian dogs, Sider said. The dogs help to keep the sheep together.

Sider's dogs have been injured by large predators, including a female dog that was slapped across the face by a bear. He does not believe wolves are attracted to his dogs in any manner, including by female dogs in heat. Sider said from his experience, wolf packs select the least-protected herd or flock to attack. Since his herds are guarded by Karakachans, and his neighbor's herd is not, it will be the neighbor who suffers.

"This livestock guarding dog has always been the only effective traditional protection against predators," according to Sider. "It is a key factor in solving the predator–man conflict and consequently saving large carnivores."

"These are the dogs which live and die as soldiers," Sider said.

Sider Sedefchev with several of his farm's working Karakachans.

Livestock Guardian Dogs

A Karakachan with a naturally short tail working in the mountains of Bulgaria.

CHAPTER 12

TURKISH LIONS

Kangals with cropped ears at work in a field in Turkey. The dogs are called "Turkish lions."

The dogs were massive. Tan-bodied and black-faced, they were lion-like, their deep barks booming in warning. Looking like something from the days of the gladia-tors, the dogs wore heavy iron collars, with long sharp spikes glistening in the sun. We had encroached on the field of grain stub-ble where the sheep flock grazed, and their guardian dogs—sometimes called Turkish lions—were responding. Lunging forward in aggression, they were headed for us.

Fortunately, we were in the company of their owner, and when we stepped closer toward him, their aggression disappeared and they greeted him with joy. Thus was our first meeting with Kangal guardian dogs in Turkey.

As word of the effectiveness of livestock protection dogs spreads among American livestock producers, so has the number of ranches that have begun using the dogs, eventually working in other guardian dog breeds from throughout Europe and Asia in the process.

There are several guardian breeds native to Turkey but growing in popularity in other countries, including the United States, but once again the politics of naming breeds is involved. Although the American Kennel Club recognizes the Anatolian Shepherd as a distinct breed, the same organization does not recognize the Akbash. We traveled to Turkey to view working livestock guardian dogs and tend to side with the dog experts in that country who classify Turkish shepherd dogs into one broad group while recognizing three distinct breeds:

- Akbash (Turkish officials spell the breed name without the "h" on the end) is the Turkish counterpart of the large white guardian dog breeds found throughout Europe, including the Great Pyrenees, Greek Sheepdog, Hungary's Kuvasz, the Polish Tatra, and the Italian Maremma;
- Kangal, a powerful fawn-colored dog with a black mask; and
- Kars, a double-coated guardian dog, that is similar to the Caucasian Ovcharka found in the republics of Georgia and Armenia.

Some dog fanciers contend there are other guardian breeds in Turkey, but we haven't found their arguments particularly compelling. There are many guardian dogs on the streets and in the villages of Turkey, but these animals seem to be variations or mixtures of the breeds listed above. Ilker Ünlü of Turkey has written extensively about the diversity of guardian dog breeds in his country, and takes issue with breed descriptions from outside that country that group all these Turkish dogs into one "Anatolian Shepherd Dog" breed.

Ünlü describes how the colors of the guardian dogs match the colors of their sheep flocks, which have fairly distinct grazing ranges in this huge country. "This is something that shepherds must have discovered thousands of years ago," Ünlü wrote. "A dog, well camouflaged in the flock, will also provide good protection by giving very little indication of where the protection is in relation to the sheep. From a distance it is very difficult to spot the dog, and just as difficult to plan an attack."[25]

The bright white Akbash dogs share the same coloration as the Kivircik sheep and Angora goats. Of the Kangal dog, Ünlü writes, "Their resemblance to the breed of sheep they protect is striking. The Akkaraman sheep with its characteristic black face and ears seem to require no other breed of dogs than a Kangal dog."

The Turkish Kennel Club also recognizes two other guardian dogs breeds: the previously mentioned Kars dog of northeastern Turkey; and the Turkish Mastiff, a newly recognized breed that appears to be more commonly used for dog fighting and estate protection than for livestock guardian duties.

There are numerous other guardian dog breeds called by local names throughout

Kangal female guarding her flock in Turkey.

their countries of origin throughout Europe and Asia. The Central Asian Ovcharkas that we use on our range in Wyoming are essentially the same *molosser* (massive) dogs known by their local breed names in various regions: Aziat in Russia; Kuchi in Afghanistan; Alabai, Kopek, and Volkodav in Turkmenistan; and Tobet in Kazakhstan.

Our travels across Turkey took us through a variety of landscape and vegetation types, from the luxurious Mediterranean seaside, to the plateaus and flats of the

Livestock Guardian Dogs

Two young guardians in a village of eastern Turkey.

interior, and to stunning mountain ranges. Guvener Isik, author of *The Sheepdogs of Anatolia* who also runs a goat creamery in a small Turkish village, served as our guide. We saw a wide variety of large guardian-type dogs throughout Turkey, and were struck by how common the dogs are—in villages, along roads, with livestock—in essence filling the large canine niche. We were told, "Where the dogs are, the wolves cannot be." Some were fighting dogs, or good herd guardians, while others were village dogs, and street dogs. The dogs were large and looked like typical livestock guardians, even if they weren't all involved in protecting flocks. We learned that in many areas of Turkey, stray dogs are collected, vaccinated, and neutered before

being released back on the streets. Villagers care for these street dogs, which of course show no aggression to humans or livestock (or else they would be quickly dispatched).

Turkey is a huge landmass, with many people there involved in agriculture. There are few fences, so livestock must be herded (most often on foot, or sometimes astride a donkey). The country has a substantial wild boar population, which many livestock producers view as a much bigger threat to their livelihoods than wolves. We saw and heard bells on sheep, goats, and dogs. We saw colorful beaded collars placed on sheep to celebrate their beauty.

Turkey's wolf population is estimated to be from five thousand to seven thousand. Wolves are considered an unprotected pest

Turkish Lions

This brindle-colored yearling male dog was guarding a herd of goats in the hills of the Bagkonak area of Turkey.

Livestock Guardian Dogs

A guardian dog crosses a creek in a village in the Mus region.

species with no established quotas for wolf hunting, or set hunting season, so they can be killed at any time. Typical wolf packs can be as large as up to thirteen members, preying on red deer, roe deer, wild boar, brown hares, and livestock. There is no compensation for damages caused by wolf attacks on livestock. There are occasional attacks on humans by wolves, especially in eastern Turkey, but there is no formal system for recording or tracking such incidents.

Isik told us that villagers believe that the behavior of wolves has changed over the years, and now the wolves don't howl, or don't form big packs very often. Most packs usually consist of only two to four animals. This is similar to what we learned about wolf behavior in Spain.

Isik told us that wolves do breed with guardian dogs on occasion, and we heard local stories about it while we were in Turkey. One village in eastern Turkey is known for its wolf–dog crosses, although we did not travel there because it wasn't part of our area of inquiry. The village is located near the Republic of Georgia where researchers were recently surprised to learn of the apparent frequent breeding between guardian dogs and wolves.

We were told a story of a shepherd who captured a wolf pup and raised it with his herd at least until it was four years old. This story was repeated to us by several different people. Other wolves in Turkey are known to prey on dogs as a food source.

We saw tethered guardian dogs and learned that these dogs are set free at night,

A guardian dog protecting a herd of cattle grazing the highlands in the Mus region.

to guard a cattle dairy from possible theft by humans. We found that generally, dogs that are tethered during the day are more dangerous to humans. Dog fighting/wrestling is still popular in parts of Turkey, and those who raise big dogs often meet up for competition between their dogs and those from other villages. We met Turkish dogs that are proven fighting dogs as well as excellent guardians.

Livestock guardian dogs in Turkey must defend their charges against jackals, wolves, leopards, and wild boars. We did not learn of any human aggression or livestock aggression in street dogs, village dogs, etc., although they are found throughout the country. We entered a mountain camp where the guardians live with the cattle on the mountain full time, but we also visited villages where cattle are penned at night, and let out to graze the highlands during the day, their guardian dogs constantly at their side.

The Kangal is now the national dog of Turkey, leading to show-dog syndrome, where some of the top dogs are called Kangals, yet have little guardian instinct, and breeders expect to be paid top dollar for pups. Soon after we arrived in the country, we realized there are several different ideas about what constitutes a "Kangal," so we stopped using that breed name when we talked to people. We learned that there are a wide variety of large shepherd dogs

Livestock Guardian Dogs

Livestock guardian dogs near a village in eastern Turkey.

in Turkey, and some are given regional names.

The Turkish Kangal is a large guardian that is able to fight wolves, and rather than just try to deter the predators, Kangals reportedly prefer to kill them. Kangals are famous for their fierce battles with predators, and many adult dogs in their country of origin carry battle scars. Yet Kangals are raised in villages with children and small animals, and are known for their steady temperament and gentle manner.

We met up with Ibrahim Kayış at his home in the village of Seyit in southwestern Turkey, and then traveled out to his fields to meet his family's sheep flock being tended to by his brother Musa, and father Hasan. Their four hundred sheep produce both meat and milk, and they've raised livestock

protection dogs all their lives, never losing a dog to a wolf in forty-five years. The guardian dogs raised by this family are famous for their massive size, standing thirty-two to thirty-four inches tall. These are long-distance dogs, traveling six months of the year, so their paws get worn and bleed on occasion.

Their flock members give birth out in the fields, and the dogs clean up the afterbirth. There is an abundance of wolves in the area, in addition to jackals, but no bears in this region of the country. Wild boars are common, and we watched a cell phone video of two of the Kayış dogs killing a 450-pound wild boar—an impressive accomplishment. Although jackals are not frequently killed because they will eat wild boar piglets, the men had recently shot two jackals in the

Four-year-old male Kangal with owner İbrahim Kayış of the Denizli area.

Livestock Guardian Dogs

village. Big wolf packs were more common a generation ago, but have declined since then, and now are making somewhat of a comeback in this region of Turkey, according to Ibrahim. A pack of seventeen wolves was roaming the area, and had reportedly crossed the roadway near where his flock grazed.

"The wolves in this area are the size of the dogs, and sometimes larger," Isik interpreted for Ibrahim. The Kayış pups are handled and played with by children, but are not petted by adults. We were told that too much touching makes the dogs "too soft."

Many of the guardian dogs we encountered in Turkey had cropped ears, and we learned most have their ears cropped when just a few days old. While Ibrahim doesn't like the look of hanging ears, his father Hasan explained it has a utilitarian purpose. "It looks nice, but the main reason apart from that, different dogs of different herds attack each other, and they damage each other, plus the wolf does the same thing, grabs the ear and damages the ear," he said. "With the wolf, he has no chance, with those ears."

The Kayis family doesn't dock the tails of their dogs, as they believe the tail adds balance to the dog, and allows them to cover their faces when it's cold. The family feeds their dogs well so that the dogs are able to challenge wolves. The dogs are fed *yal*—a grain-based porridge, in addition to wild boar and dead sheep. Ibrahim noted, "Just because you feed him a dead sheep doesn't mean that he'll eat a live sheep."

A yearling male Kangal owned by the Kayış family, demonstrating the dog's double back dewclaws.

The Kayış family hadn't experienced any recent losses to wolves, and they keep three or four dogs with four hundred sheep. They said this combination will take care of the wolves, but there will be some lambs lost, even though the sheep are penned at night.

Ibrahim suggested the reason so many American livestock guardian dogs are killed by wolves is that their genetics may be getting diluted, or getting "softer" genes passed, or by crossing with other lineages lacking enough desire.

The men admitted that their dogs can be aggressive toward humans, and several people have climbed trees to get away from their dogs. They alleged that it depends on the people—that the dogs will chase and attack strangers. People in Turkey will try to steal sheep, we were told, and so the dogs need to be aggressive to humans at times. The dogs independently judge the human and its activities, and respond in the manner it deems appropriate. In Turkey, if a dog bites a person, there is no real recourse for the human victim.

The day before our visit, two flocks crossed paths, and the neighboring sheepherder stopped to pet one of the Kayış dogs. All was well with that, but then one of the neighbor's sheep joined the Kayış flock, and when the neighboring herder attempted to grab his sheep back, he was grabbed on the buttock by the very dog he had just petted. The herder did not say anything bad about the situation, realizing he should have known that's what would happen.

"The people need to be educated, instead of changing the dog, people need to

The author with a young male Kangal.

change," Isik said. "These dogs have a job to do and they do it."

"The man should be wiser," Isik said. "If your dog has no aggression, how is it going to guard your sheep?" Isik asked. They test their dogs all the time, Ibrahim said, and he sells dogs with a guarantee that the buyer won't have livestock losses.

The dogs work together as a pack. When danger approaches, one stays in the herd, while the others surround the flock. The dogs are affectionate with the sheep, and do not kill the family's chickens, which range freely around the dogs. One of their dogs lived to be thirteen years old, but the average lifespan is seven years.

As we interviewed members of the Kayis family in the unfenced fields where their flock grazed, the mutual affection between the men and their animals was apparent. The dogs greeted the men with enthusiasm, and a few pampered sheep (with wool dyed into vibrant colors and wearing colorful beaded collars) approached them for attention and petting. As the sheep flock grazed near the border of the grazing area, the men were able to verbally call the sheep away from forbidden range, an indication of the close working relationship between these human and their animals. We later traveled over a bridge spanning the wide Büyük Menderes River, the site of a local festival each fall in which a competition is held for shepherds. The competition involves the shepherd jumping into the water, calling his flock to join him in a swim across the river. The shepherd with the fastest time is the winner. We were not

Kangal dogs at work in Turkey.

A Kangal female at work in the fields of Turkey.

surprised to learn that Musa Kayış was the reigning champion.

We met up with a cattleman in a small village in southeastern Anatolia who grazes about one hundred cattle with one or two guardian dogs, in addition to a herd of four hundred Mandak sheep. The cattle drovers have very close contact with their herd members throughout the day. This region has a four-month winter where the cattle are free outside the village, while the village sheep are penned at night in rock barns with windows. The windows allow for airflow, but wolves occasionally use them to gain entrance and kill sheep. The villagers we interviewed all feed *yal* to their dogs, dock one or both ears, and feed dead animals to the dogs as well.

We met goat producer Ali Keskin near the village Selcen as he stood over a billy goat trying to figure out why it was suffering from paralysis. Ali uses three dogs with 250 goats, including a two-and-a-half-year-old unmanageable dog that had pulled two people down from tractors. This obviously dangerous dog was still alive only because it had killed a wolf that came into his barn

Livestock Guardian Dogs

a month prior, but Ali was willing to offer him for sale to us. We declined.

Ali does not pet, socialize with, or train his dogs in any way. Ali docks his dog's ears because the dogs fight with wolves. He said docking minimizes the damage done to the dog, and because "A dog with ears cannot hear." Ali doesn't like to use spiked collars in brushy country because the collars can get stuck, he said. Wolves had killed some of his dogs in the past. Now that wolves are congregating in bigger packs, he said he has constant wolf problems.

Father and son Hayuk and Ilhan Kultu are Anatolian Turkmen and sheep producers. They had four sheep killed by wolves, so they brought in a new bitch—a wolf-proven bitch. Their dogs have been injured in fights with wolves, without loss of sheep.

They also have a dog that fought and killed a bear. They recommend eight dogs as the proper number to protect a thousand sheep, or three or four dogs with five hundred sheep. This family does not favor dogs called Kangals—their admiration is reserved for the guardians called Central Anatolians. These dogs can kill badgers, they said, and not many animals can.

The dogs fight between themselves, the Kultus reported, but noted that a good dog manages the rest of the dog pack, and the lead dogs are usually bitches. In pursuit of wolves, the Kultus don't like their dogs to go too far.

The men explained that in 2010, every flock in their region suffered losses from wolves, with four or five sheep lost from each flock (with flock sizes varying from

Typical guardian dogs in the eastern portion of Turkey.

A typical Anatolian-type dog found across Turkey.

five hundred to one thousand animals). They also believe that wolf behavior is changing—that the wolves are becoming more shy and sneaky.

The Kultus dock their dog's ears, but not their tails. They recommend always cropping the ears, with no exception, because when the dogs have conflicts with wolves, they lose their ears. They don't crop ears of pups born during the summer months, but when those dogs grow up, they get in conflicts with wolves and get their ears ripped. It was also noted that the long iron spikes used on the traditional collars can damage uncropped ears. Some

of their dogs have naturally docked tails, and the men prefer dogs with back dew-claws. They do not castrate nor do they neuter any of their dogs. Hayuk said that the size of the dog doesn't matter; rather, "It's all about the heart." He admitted that his bear-killer dog was a very big dog. Big dogs are not good for rocky areas, both men said, adding that they select for hard, compact paws.

As we drove across the country, we regularly saw dead dogs (large livestock protection dogs or herding dogs) that had been killed by vehicle collisions. Rural villages have chickens and guardian dogs on

A long-haired Kars guardian dog on the rangeland of eastern Turkey. Kars dogs are similar to the Caucasian Ovcharka found across the Turkish border in the neighboring republics of Georgia and Armenia.

the streets, but many communities are losing their people and their livestock herds. We were told they need young people, especially young women to become wives for the men, because all the younger people are moving to the cities. We were told that the number of men, dogs, and cattle is decreasing across the countryside.

My friend Patrick Porter of Massachusetts is a farmer and he is also one of the most talented writers I know. In the spring of 2010, he sent me a letter describing meeting a Turkish man, a server at a restaurant some five years prior. Patrick's son Liam asked the man what the dogs are like in his native land, and Patrick graciously gave me permission to share the story.

"We are fortunate enough to have the Kangal. My father keeps four of them. They protect our sheep and saved the life of my sister from wolves three years ago.

"Nowhere in the world will you find a better animal . . . I have been in many countries and there is not a Kangal anywhere. That is why the world will be at war soon. If everyone had my father's dogs, things would already be settled. In my home . . . the wolves will come for something that is not theirs. The Kangal finishes it all and sends them away. We are herders, and we like it. The Kangal keeps us safe in our land. We cut off his ears when he is little, so the wolf cannot bite. The wolf is always killed by the time we get to the dogs. My sister, she has two sons now. The wolves came for her and were destroyed . . . now she has two sons."

Liam asked if the Kangal is a good dog for a family, and the man replied: "Sir, the Kangal is good for the world. Not just families. They are the color of sand and rock, they have the heart of those before them, the wild animals that have become our friends. It is that way with our dogs. Everyone should have a Kangal."

A guardian dog in the mountains of Turkey.

Livestock Guardian Dogs

CHAPTER 13

SURVIVAL TOOLS

A livestock guardian dog wearing a traditional spiked collar in Turkey.

As Jim and I traveled through rugged landscapes of Europe and Turkey, we encountered flock after flock of sheep, goats, and cattle. Massive guardian dogs accompanied all the flocks, most wearing spiked collars. The spiked collars are an ancient tool used by livestock producers throughout the Old World to give livestock guardian dogs an advantage in combat with wolves, and are still commonly used for that purpose today. The collars not only provide a defense for the dogs against wolf attack, but are also used as weapons by the dogs in such encounters.

Shepherds in Turkey do not kill wolves—they expect their dogs to handle the threat posed by their wild canid counterparts. "If you have good dogs, the dogs will take care of the wolves," Ibrahim Kayış said. To shoot at a wolf would be an admission that the guardian dogs were not capable of doing the job they were meant to do. "If you trust the dogs, they can handle them."

While the dogs need to have the "heart" to challenge wolves, their shepherds do not send the dogs into battle without proper equipment. The men said that spiked collars are necessary to give the dogs an advantage against wolves when they engage in battle. If the dog is wearing a spiked collar, the wolf cannot grab the dog around the throat or the back of the neck, Hasan Kayış explained, and the dog will often hit the wolf with the spiked tips on the collars to inflict damage. Older, more experienced dogs learn to use the collar as a weapon, Hasan said: "As they get older, they know the use of it [the collar], and they start actively using it."

Because the collars can inflict damage on other dogs, they are generally only used on the most valued or aggressive, mature dogs. These are the dogs most likely to give chase to a wolf, and the first to challenge a wolf pack.

These Turkish sheepmen explained that their guardian dogs and wild wolves communicate with each other, just as guardian dogs associated with different livestock flocks communicate when they pass through the same vicinity. As the herds make seasonal movements for grazing, the dogs move as well, and the dogs from different flocks come into contact, and sometimes into conflict. But much communication is verbal.

"The male dogs are very dominant," Hasan said of the guardian dogs. "The way they howl tells the coming dogs to stay away, by the tone of their voice. But some dogs, they want to challenge, so they come in anyway."

"The wolf also knows the mighty dog," said Hasan. "The wolf can feel that these dogs are dangerous, so it does not approach. It's the same, with the dog and the wolf—the voice tells the wolf its strength."

We were told a typical wolf encounter involves one wolf's initial approach toward the flock. If the dogs do not react too aggressively, the entire pack of wolves then attacks. Some encounters with wolves take place over long times and distances. Sometimes the dogs will go for several miles in chase. Hassan explained that the chasers are the wolf killers—sometimes not returning for several days.

Turkish shepherds Hayuk and Ilhan Kultu have had their dogs injured in fights with wolves, without a loss of sheep in the process. All of their dogs wear spiked collars except when around the village so the dogs aren't given an unfair advantage with village dogs. The men explained that the iron collars are too cold in the winter, so thick felt is used to line the collars and make them more comfortable for the dog. Hayuk agreed with his other shepherd counterparts about the dogs not just wearing the collars in battle, but also actively using them as weapons.

"A smart dog knows how to use its spiked collar," Hayuk said.

Our travels took us to a small Yoruk village in Turkey's Central Anatolia region where we met with Memet, a local cattleman. Memet said his area is "infested' with wolves, so guardian dogs must stay with the village herd at all times. If the cattle are

Ahmet Kemal sews a felt lining to an iron collar in his shop in Konya, Turkey.

Survival Tools

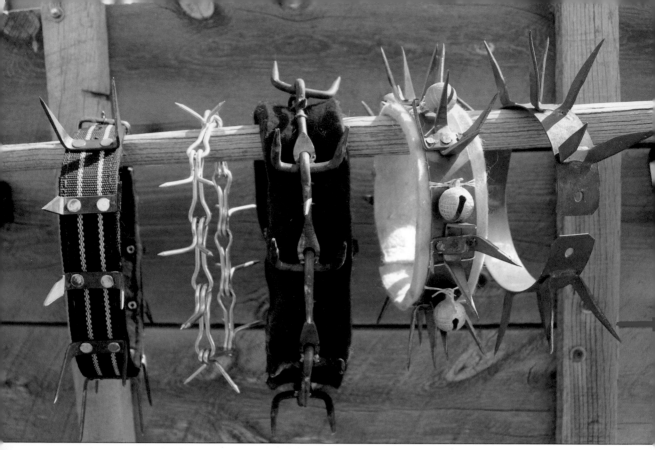

These are a series of spiked collars used in Turkey. Iron collars can be manufactured by any skilled blacksmith. The length of spikes varies. These collars are made to protect the dog's neck from wolf bites when in battle.

related and have close bonds, they do well, he said. During an attack, Memet said, the cows will bunch into a protective circle around the calves, and normally wolves are unable to breach the circle to make a kill. He said that some of the cows would also defend their livestock protection dogs. There is very close cooperation between the animals, Memet notes.

Village shepherds had recently met and decided not to use spiked collars on their dogs because village dogs have been killed in conflicts with guardian dogs wearing these collars. We learned of similar conflicts in areas where protection dogs came in contact with village dogs.

But banning the use of spiked collars had an unintended consequence, as Memet explained. Without the protective collars, livestock producers were losing more dogs to wolves, so they were using more guardian dogs, and they were experiencing much more fighting between the dogs.

We heard the sound of a shotgun being fired while we were at a cattle camp in the mountains of Turkey late one evening. We were told herders were shooting to keep the wolves away, but they didn't want to actually shoot the wolves because wolves control the wild boar population. To these livestock producers, wild boars are responsible for much more damage than the wolves.

Livestock Guardian Dogs

This is a series of traditional Bulgarian spiked collars, showing the simple, single-piece collars cut from one piece of strap metal (cut from old saw blade).

Livestock producers using iron collars on their guardian dogs in Bulgaria offered some of the same views as the shepherds we met in Turkey. Sider Sedefchev, whose herd of native Karakachan sheep graze high in Bulgaria's Pirin Mountains, said the spiked collars are an effective tool for protecting dogs that become involved in physical conflicts with wolves.

"They can use this," Sider said. "The more experience they have, they know how to use them. Our best dog would use his, he was a professional fighter. He liked to fight but was very intelligent."

Sider explained that if the dogs are working well together, the spiked collars serve as effective protection for the dogs. But if there is conflict among the group of dogs, Sider cautioned that the collars could cause additional problems, such as injuries inflicted during fights. Once a new dog finds its place in the pack (after the hierarchy fighting is over) he places iron collars on the dogs.

Sider said he placed spiked collars on the three bravest dogs guarding his flock because these will be the dogs to immediately challenge the wolves. This strategy provides a higher likelihood that the dogs will survive the initial attack while the other, less aggressive dogs are coming to join in, he explained.

When we visited, Sider wasn't currently using spiked collars, but has used them in the past, and planned to again in the future. He cautioned that dog packs sometimes have inter-pack strife and iron collars can result in broken teeth. He reported the collars do not cause problems in heavy brush, or getting snagged.

The livestock producers we met in central Spain's wolf country also used spiked

A traditional collar on a dog in Spain, featuring rows of four spikes. The collar is made of two thick leather straps, riveted together to protect the spike heads.

collars on their Spanish mastiff dogs, but their collars were made of leather rather than iron. Most of the livestock producers we met in the central Spain did not have spiked collars (called *carlancas* in that country) on their dogs, although *carlanca* use is predominantly a practice of northern Spain. In higher wolf density areas, or with longer recent history of wolves, *carlanca* use is more common.

We visited the Caserio De La Torre, a large range cattle operation that breeds Avileña cattle, a native Iberian breed known for its bravery. Carlos, the ranch foreman, explained that a pack of wolves lives on the ranch, and constantly challenges the ranch's Spanish mastiffs, so the dogs must wear their spiked collars.

Carlos recommends using iron collars in summer, and switching to leather in winter (since the leather collars will sweat in summer). He cautioned that the dogs sometimes try to scratch their heads when the collars are first fitted, and are injured that way. Carlos recommends the collars be used year-round where wolves are present.

Robert Vartanyan of Russia wrote that Caucasian Ovcharkas in the Transcaucasia region wear metal-cast collars with spikes. The collars were formed from old hand-

Livestock Guardian Dogs

held sheep shears. The use of spiked, anti-wolf collars on guardians has been reported from various regions of the world, including Italy, Poland, Romania, Spain, Turkey, Bulgaria, and Portugal. Use of spiked collars has been very limited in the United States, based on a variety of factors, including questions of how to properly use them; concern about the safety of the collars where there are barbed or woven wire fences and heavy brush, and/or frigid temperatures and snow; and the lack of access to the collars.

There are two general types of spiked collars: leather or fabric, and iron. Leather or fabric collars may be a safe alternative during the winter months when herders are concerned about harm to the dogs from having iron collars on their dog's necks with snow and cold temperatures. The heavy iron collars, if manufactured in a large enough size to drape over the lower portion of a guardian dog's neck, may provide enough room for the dog to slide the collar over its head should it become hung up in a fence or brush. Although herders live with the dogs in range-sheep outfits, the dogs are not always in sight and may only be seen once or twice a day. In this situation, ensuring the safety of the dogs wearing the collars is an important consideration. We use the leather version of the spiked collars and fit them to the dog, so that it can slip out of the collar if it gets stuck in a fence. Better to lose the collar than lose the dog.

More livestock producers in the United States are beginning to use the spiked collars on their dogs. One Montana producer credits the use of a spiked iron collar with saving the life of one of his guardian dogs. When the guardian dog rushed out to challenge a wolf that was approaching the flock, the wolf and dog engaged in battle. The wolf quickly retreated with a bloody mouth, and the dog sustained only minor injuries. The dog was treated on site and continued its work.

As livestock producers in the United States react to expanding populations of gray wolves in the Northern Rockies, Southwest, and Great Lakes regions, they too are turning to the use of these archaic tools, proving that advances mustn't always be based on modern technology. Ancient tools can be used to meet current challenges.

Turkmen strap a plain leather collar with bells over the top of a traditional spiked collar that has been lined with wool felting to keep the guardian comfortable in cold temperatures.

Livestock Guardian Dogs

CHAPTER 14

PROBLEMS AND BENEFITS

The dogs become accustomed to wild animals they do not perceive as a threat to the herd.

Individual guardian dogs, regardless of breed, have different behavioral tendencies, and a dog that might work well for one ranch may not meet expectations on another. Small farm owners will not want a guardian that roams a large range while it patrols, and migratory sheep flock owners will want a dog that is agile enough to migrate with the sheep. The use of guardian dogs means that the use of traps, snares, and poisons is prohibited in the dog's territory because of the potential of harm to the dogs.

Livestock producers may experience a variety of problem behaviors with their livestock protection dogs, including chasing wildlife, aggressive play or harassment of herd members, aggression toward unknown people, and roaming from the ranch. Adolescent dogs can be obnoxious to sheep at times, trying to initiate play, pulling wool, or chewing on ears of the flock members they are supposed to guard. An attentive shepherd reprimands the dogs for such behavior, and most dogs quickly grow out of the problem behavior after correction. But the dogs mature slowly, and will make mistakes that need correction.

Other issues can include barking (it's in their job description), and digging and chewing objects (after all, they are dogs). The dogs will encounter and carry porcu-pine quills, ticks, fleas, and internal worms, and will be sprayed by skunks.

Basic training needs to include a few verbal commands at a minimum, including "No," "Come," and "Go to the sheep." The dogs should be taught to walk on a leash, be tethered, and kenneled.

Guardian dogs cost money to purchase and require some level of training and supervision, making them a long-term commitment. Some will get sick, hurt, or stolen, and they all die at some point. Most guardian dogs have relatively short life spans (about ten years) under the best of conditions, and many never make it this long as they are killed in accidents, hit by vehicles, or are shot by neighbors, hunters, or those who do not understand their purpose or believe they are a threat to game animals.

With multiple guardian dogs, fights will occur on occasion.

Dogs fight among themselves, yet they continue to live in the same system, in the same environment, and will see each other numerous times in the future. Very rarely are the dogs killed in such fights. Even females will fight.

Unless specifically bred and trained otherwise, most livestock guardian dogs pose little threat to humans, yet will greet strangers approaching their herds with a show of aggression, mainly by running forward with excited barking, which serves as a warning. Most guardians will behave more aggressively toward people greeting them with unacceptable behavior, such as hysterical screaming, erratic drunkenness (yes, this is an issue in some areas of the world), or people who attempt to continue through the flock despite warnings from the dogs.

Range sheep flocks use lands also enjoyed by recreationists, including bikers, hikers, and backpackers. When those recreationists get too close to the sheep flocks, the dogs will confront them, very rarely attacking, but behaving aggressively enough that people are intimidated to the point of avoiding the flock.

In a widely publicized case in 2008, a woman involved in a Colorado bicycle race was injured by two guardian dogs and later sued the dogs' owner, who later had the dogs euthanized. According to press accounts, the woman had quickly, quietly, and unexpectedly raced her way at sunset into a sheep flock that was grazing on national forest land, and the guardian dogs swiftly attacked, tackling the woman on the bike. The flock's owner had not been notified by forest service officials that the area would be swarmed with mountain bike racers that weekend, and thus had not been given the opportunity to move his herd and

dogs out of an area of potential conflict. It was a tragedy for all involved.

We post signs noting the presence of guardian dogs and sheep at the entrance to our pastures, and hand out flyers to neighbors adjacent to the pastures where our flock grazes seasonally. Letters to the editor of area newspapers, or press releases, can be helpful in educating recreationists in grazing areas about how and why guardian dogs are used, and how to behave around them.

While recreationists on all-terrain vehicles, or four-wheelers, can often be heard long before they approach, a rider crouched over the handlebars of a mountain bike can be startling to the sheep and dogs, and often won't be heard until the distance is very close. Other conflicts arise from off-leash dogs accompanying recreationists, including hunting dogs. If any of these dogs approach a grazing herd, the guardian dogs will react. Recreationists should be encouraged to make a wide detour around grazing or resting herds, and should be thanked for doing so.

Guardian dogs have varied levels of aggression, and this in itself can prove problematic. Some dogs are too aggressive, but not toward humans: we knew a female guardian dog that would kill any pup that wasn't hers. Other guardians will chase big game animals such as deer, or will prey on their fawns in the spring. Some dogs may not be aggressive enough, such as our Tucker that would confront coyotes and hold them back from the herd, but wouldn't bite or kill them. Some dogs are so friendly that they approach strangers, only to be stolen or "rescued" by someone wanting a big beautiful housedog. But these dogs haven't been bred or raised to be indoor pets, so they often end up in shelters because of behavior problems. The usual problem is

Minimize conflicts

To minimize conflicts with guardian dogs, USDA Wildlife Services recommends recreationists:

Do:
- Stop and dismount if mountain biking.
- Put your bike between you and the dog.
- Walk your bike until you are well past the sheep.
- Keep your distance from the flock.
- Keep your dog(s) leashed at all times.
- Watch for the protection dog(s) and the herder.
- Remain calm and quiet if a dog(s) approaches.

Don't:
- Hike or ride your all-terrain vehicle/mountain bike into or near the flock.
- Make quick or aggressive movements around the dog(s) or sheep.
- Attempt to hit or throw things at the dog(s) or sheep.
- Yell at dog(s) or sheep, unless approached— then, yell "go back" or "no!"
- Try to outrun the dog(s).

Signs like this one can be modified with grazing dates and contact information.

Benefits

Jeffrey Green and Roger A. Woodruff of the U.S. Sheep Experiment Station listed numerous benefits from the use of livestock guardian dogs, including:

- Reduction of predator losses, causing an increase in net profits;
- Reduced labor;
- Pastures more efficiently utilized, increasing the condition of the sheep;
- Dogs alerting the owners to disturbances;
- Opportunity to increase the size of the flock;
- Less reliance on less desirable methods of predator control;
- Protection for family members and property; and
- Peace of mind.

Livestock Guardian Dogs

that the dog simply needs to be returned to its livestock flock.

Guardian dogs have to be fed every day, or nearly every day. They do require supervision and there is a learning curve associated with establishing guardian dogs on your ranch (for you, your herd, and your dogs).

We've noticed that when something frightens our sheep, they will come to the dog for protection. Idaho graduate student Bryson Webber studied sheep movements, finding that the presence of guardian dogs tends to relax the sheep, allowing them to travel greater distances, and resulting in more effective use of pasture resources, as well as weight gains.[26]

In addition to the very important benefit of reducing depredation on livestock herds, our guardian dogs provide other important advantages that are often overlooked:

- Cleaning up afterbirth, which results in a cleaner pasture that is less desirable to predators.
- Reacting negatively to the presence of ravens and other large birds, so weak newborn livestock are provided with protection from avian predators as well.
- Maintaining a separation of our flock from specific wildlife populations.
- Possibly saving human lives, or saving you or your family members from injury.

Rant, my Central Asian Ovcharka, proved this last point to me when he continuously knocked me down and kept me from getting through a stand of willows where a bear was killing sheep on the other side. At the time, I had no idea a bear was present, or why the dog was behaving so aggressively toward me (without harming me). In some areas of the world, large predators such as tigers, leopards, and wolves view humans as potential prey, and

livestock guardian dogs serve as an early-warning system, alerting residents to predator presence, thus reducing the potential for loss of human life.

Kurt VerCauteren and Thomas Gehring are among a small group of researchers who are looking at guardian dogs in a new way in recent years, documenting other benefits from their use.

In research in Michigan, VerCauteren and Gehring and their study co-authors found that once livestock guardian dogs were placed on cattle farms, visitation to those farms by coyotes and wolves was nonexistent, but increased on the control farm where no dogs were present, and although no livestock depredations occurred on farms where guardian dogs were used, neighboring unprotected farms did experience depredations. An additional benefit was to ground-nesting birds on cattle farms where the guardian dogs were present. Birds and their nests were found on the farms using livestock guardians, while no nests were found on the control farm. The researchers suggest that the dogs reduced pasture visitation and nest depredation by mesopredators such as skunks, opossums, and raccoons.

The study recommended the use of livestock guardian dogs "as a proactive management tool that producers can implement to minimize the threat of livestock depredations and transmission of disease from wildlife to livestock. Livestock protection dogs should be investigated further as a more general conservation tool for protecting valuable wildlife, such as ground-nesting birds that use livestock pastures and are impacted by predators that use these pastures."[27]

VerCauteren also conducted an experiment using livestock guardians to deter potentially infectious deer from livestock pastures, and to prevent deer from consuming livestock feed (a primary route to transmission of tuberculosis from deer to cattle). The study concluded that the dogs may reduce the potential for disease transmission between deer and cattle by reducing deer use and time spent in pastures. His research also found a decrease, or even total elimination, of damage by wild boar (also a host species for tuberculosis) in areas used by livestock guardian dogs.

The potential use of guardian dogs to reduce disease-transmission risk is rather broad, from brucellosis transmission to cattle from elk and bison in the Yellowstone area, to keeping bighorn sheep and domestic sheep separated to reduce the possibility of pathogen transmission that may result in pneumonia in the bighorns. Western ranchers already provide anecdotal accounts of using guardian dogs to keep wild and domestic sheep separated, as well as guardian dogs keeping moose and other animals out of stored feed. Texas cattle ranchers use guardian dogs to limit the potential transmission of cattle tick fever from whitetail deer to beef cattle.

We encounter greater sage-grouse broods with greater frequency and in larger numbers in areas where our guardian dogs are working. The simple fact is that the dogs reduce the presence of other predators, resulting in greater chick survival. Guardian dogs can be used as a tool to increase game bird production. If guardian dogs do not disturb game birds, the dogs may be used to reduce predation on these desirable species, and to improve recruitment of offspring.

Pronghorn antelope are another species that appears to benefit from the presence of our guardian dogs. Every year during lambing, we notice a few pronghorn does that come into the areas where our herd is lambing and subsequently give birth there. The pronghorn seem to realize that the threat of depredation is greater outside of our domestic sheep herd. The pronghorn herds share the same range with our sheep herd year-round, and are within sight of the dogs from the time that they are puppies, so the dogs are acclimated to their presence and do not bother them.

In the winter months, we use a heated stock water tank for the sheep. Groups of pronghorn will come in to drink from the tank, and bed down near the sheep, always under the watchful eyes of the guardian dogs. The dogs can patrol around the mixed herd of wild and domestic animals, with no sign of alarm from either. During one bitterly cold winter week, nearly one thousand pronghorn stayed close to our flock and under the protection of our guardian dogs, an effective strategy on the range that was inhabited by at least one pack of five coyotes that winter.

During the late-summer pronghorn antelope breeding season, we always have a buck or two that will approach the dogs, snorting and "barking" in their direction. The dogs have amazing tolerance for these bucks, seeming to understand they are not a threat, but more of a nuisance.

Later in the fall wildlife migrations, our dogs actively work to keep large herds of migrating pronghorn separated from our flock of domestic sheep. As the herds travel through our range, the dogs will stay on the outside of the flock, serving as a buffer between the two groups.

The situation is somewhat different with mule deer. For some unknown reason, we've had a number of guardian dogs that apparently do not want mule deer near the sheep flock, and will chase them a short distance before returning to the herd. But each dog is different, and we've had both a Central

Two guardians with members of their Hereford cattle herd. These versatile dogs guard both sheep and cattle herds as the need demands.

Asian Ovcharka and an Akbash that have guarded our sheep flock on its winter range, which was shared with mule deer—a situation that could have been precarious. The dogs stayed between the deer herd and the sheep flock, keeping them separated, and did not chase the deer.

Others are using livestock guardian dogs to protect deer, with captive deer breeders using livestock guardian dogs to decrease coyote predation on fawns. Early bonding between these species sets the stage for future success.

Laurie Marker and the Cheetah Conservation Fund have boosted cheetah conservation in Namibia by placing guardian dogs with livestock producers, reducing the killing of cheetahs in that region.

Maremma guardian dogs are being used to keep wild dogs from killing livestock in Queensland, Australia. Maremmas are also used to guard an endangered little penguin colony in Australia. The pair of Maremma dogs from Italy was bonded to penguins at a young age, and now patrols the colony during breeding season, protecting the population from predations by red fox. The program has had enough success that livestock guardian dogs are now being used to protect large seabirds called Australasian Gannets.

The Carpathian Mountains of Slovakia have provided a living laboratory for the study of large carnivore conflicts and the use of guardian dogs. Brown bears, gray wolves, and Eurasian lynx inhabit the

Problems and Benefits   – 173 –

An Akbash guardian dog and a coyote in a stand-off.

region, overlapping domestic sheep grazing range by 90 percent, and major wild prey available include roe deer, red deer, and wild boar. Robin Rigg of the Slovak Wildlife Society was involved in a program to supply Slovensky Cuvac and Caucasian shepherd dogs to farms in the Carpathians. Twelve percent of the flocks involved in the study accounted for about 80 percent of all losses to bears and wolves, with the same flocks affected each year. Large carnivores affected both species. A small number of these flocks had incidents of "surplus killing," with wolves killing numerous domestic animals (more than necessary for consumption). Rigg reported that the presence

Livestock protection dogs will kill and consume small rodents within their occupied territory.

Problems and Benefits

of guardian dogs "was associated with lower levels of predation and an absence of surplus killing."[28]

Regardless of the predator the guardian must deter, or the species of animal the guardian protects, it's worth noting that if livestock guardian dogs were not effective, as well as being a joy to live with, livestock producers would not have used them for thousands of years.

If livestock guardians were not effective, they would not be utilized over such a broad expanse of the world.

CHAPTER 15

AFRICANIS

A cattle protection dog following its herder in Lesotho.

When most people think about live-stock guardian dogs in Africa, what comes to mind are the well-publicized pro-grams that have imported Anatolians and other large Eurasian guardian breeds to rural pastoralists in various African countries in order to reduce conflicts with cheetahs and other native predators. While these programs have been successful at reducing livestock depredations, there are indigenous dogs in Africa that protect livestock while not complying with the guardian dog norm: from the Azawakh in the Sahel desert, to AfriCanis in South Africa, the continent is home to numerous indigenous multi-purpose dog breeds that have guarding livestock among their duties. We met several native breeds when we traveled to southern Africa in 2012.

We entered Lesotho via the eastern route, bouncing along in four-low up the rocky and kidney-bruising Sani Pass (formerly a footpath used by the Sani), at an elevation of 9,429 feet—the highest mountain road in Southern Africa. As we traversed the switchbacks, sunlight glittered off the icicles hanging on the rock walls above, and we saw glimpses of rugged green canyons down below. The Kingdom of Lesotho is an independent nation entirely surrounded by South Africa, and is regarded as the highest country on earth, with its lowest point about 4,600 feet above sea level.

As we began climbing away from the border post and even higher than Sani Pass Top, we saw herds of wool sheep grazing the great unfenced veld. We soon encountered men and boys walking with herds of cattle, sheep, and Angora goats, or riding hardy Basotho ponies, leading their burros packed with one hundred-pound bags of maize. There are few vehicles in this mountain environment, but there are burro trails crisscrossing the rural countryside, and with access so limited, most items moving into and across this region are loaded on a burro's back. Dogs accompanied most people we encountered in the highlands, some of which were obviously livestock guardians.

A few of the bigger dogs were typical of livestock guardians found around the world. The primary predators in Lesotho are jackals, but livestock guardian dogs are used to protect domestic livestock from predators such as caracals, hyenas, cheetahs, leopards, baboons, and lions.

The Basuto people who live in this rural region of Lesotho raise livestock and have a subsistence agricultural lifestyle. They live in thatch-roofed, rock- or manure-blocked huts with thick walls providing excellent insulation. A family compound may have several of these huts for extended family members, with rock corrals at the center, where livestock and hay can be safely contained. Their dirt-packed yards are manicured to perfection by women using bundled brush as brooms. Many families raise chickens, which roam freely and are very important to each household's food security. Children carry eggs as a protein-rich snack in their school lunch. The dogs leave the chickens unmolested.

Common everyday attire includes the traditional wool blanket. Many Basuto people raise Merino-type fine-wooled sheep in this alpine zone, with communal shearing sheds busy handling flocks in September. Some of the wool is sent to South Africa for processing, and returns to village stores as wool weft blankets.

The road eventually wound through Sehonghong Gorge, where we crossed the river and went up the other side of the ridge, looking down at a few vehicles that had missed the turns in the road, tumbling into the gorge, never to be retrieved. We arrived at the Saint James Mission Guest Lodge after dark, and were greeted by the mission's two friendly Anatolian-type guardian dogs. Although most non-governmental organization programs that distribute live-

A large livestock guardian dog watching over a flock in the Lesotho highlands.

stock guardian dogs in Africa use Anatolian dogs, we saw few of these dogs in our travels in Lesotho and South Africa.

The next morning, I stepped outside in the dark in the crisp morning air, petting the mission dogs while looking at the bright stars of Orion in the winter sky. Long before it was light, around 4 a.m., the roosters began to crow, and a little later, I began to hear the noisy calls of Hadada Ibis, accompanied by the sounds of braying donkeys. As the morning sunrise cast a gentle glow (direct light wouldn't come over the mountain for a few hours), the hillsides came alive with men going out to tend livestock, and first village workers, and then mission

school students, enlivening the landscape below my perch on the hill. Various types of dogs were scattered throughout the veld.

Our group consisted of Jim and me, as well as Edith Gallant, an expert and advocate for the native AfriCanis dog, and Matthew Berry, a dog trainer and expert on Maluti livestock protection dogs. While most conservation groups have focused on importing livestock guardian dogs from Eurasia into Africa, a few programs have begun placing the Maluti (a breed indigenous to Lesotho) into areas where leopard protection is being promoted where herds of cattle, sheep, and goats range. The guardian dogs do not seem to fit a specific breed

Although AfriCanis dogs are used for hunting, they must not be aggressive toward the family's chickens or other livestock.

type, and instead exhibit great diversity in their physical characteristics. Regardless, they're effective guardians.

An added benefit of the use of livestock guardian dogs that originate in Lesotho and across the grasslands of South Africa has been protection of three crane species that use the same range (blue, grey crowned, and wattled cranes). The presence of livestock guardian dogs on the landscape has provided for the protection of these bird species that are also subject to predation by jackals, caracals, and servals.

We stopped to visit Ezekiel, a Basuto stockman we met alongside a narrow roadway as his herds of cattle, sheep, and goats crossed in front of us. With his pack of assorted dogs eagerly vying for his attention, he explained that while his larger dogs serve as livestock guardians, he was doing some active crossbreeding to run jackals for the sport of the activity. The man had hundreds of livestock in each of his three herds, all accompanied by dogs with various duties, and all devoted to their master.

We encountered a young man named Realcboha Morapchi (which he said translates to "Thanks" in English). He was accompanied by three dogs of various sizes, from a short-legged dog (which he was breeding to go after small animals such as red rock hares during hunts), to two larger

A Maluti livestock protection dog at Matthew Berry's kennel in South Africa.

dogs that are more suited to livestock guardian duties and chasing jackals. He told us that he prefers his dogs and other animals to people, and said it is his hope to catch a baby jackal and raise it. He was home on break from his school studies in the nation's capital, Maseru, where he was studying to become an accountant.

Thus the first two men we talked to in depth about their dogs explained that they were selectively breeding their dogs for certain desired outcomes (the long dogs for coursing jackals, and the short dogs for burrow hunting). It became a familiar pattern as we talked dogs with African people.

We drove and visited with Basuto people throughout a large area near the main village of St. James, nestled in the flowing Maloti Mountains. We shared the potholed road with people walking, often herding

A Lesotho boy with his canine companions.

Livestock Guardian Dogs

A mated pair of guardian dogs in Lesotho.

animals. We saw large dogs guarding live-stock herds, and watched as one young boy approached a flock, only to be chased away by the guardians. The herders lounging nearby in the early morning sunshine shouted to the dogs, and they immediately returned to the herd. We witnessed a similar event involving another child who had apparently entered a forbidden area.

Although we did see a few dogs that we would classify as typical livestock guardians, most of the dog population we observed in Lesotho appeared to be smaller, multi-purpose dogs, suited and adapted to the region in which they live. Some dogs that are used for guardian duty also serve as hunting partners, and their owners were partaking in active breeding and cross-breeding programs according to personal preference, most often based on utilitarian values.

As we explored the heart of traditional Zululand in South Africa, we noticed many people carry a unique walking stick or club as they walked. The *knobkierrie* is an African club that is shaped like a walking stick with a heavy knob. A weapon of war, or of defense, the *knobkierrie* can be used to dispatch an animal brought to a stop on a hunt, or thrown as a missile to hit an animal at a distance. *Knobkierries* are widely carried in rural areas of southern Africa, and even have a place of honor on the Lesotho coat of arms. It's an effective tool for shepherds and others who do not have ready access to firearms.

We crossed the river and entered the Valley of the Tugela. The Tugela River origi-

Lesotho dogs at a homestead.

nates in the high Drakensberg Mountains, and ends as the waters flow into the Indian Ocean. We followed a road along the river, looking at herons and egrets as we visited villages and homesteads to view their dogs.

The dogs of Zululand are called Afri-Canis, meaning the native land race of dogs from southern Africa. They are medium-sized and appear to have a variety of uses (including both as hunting dogs and guardians) as the situation dictates. Although the dogs were numerous, we did not see any that were obviously village or stray dogs—the dogs we saw clearly had owners. When we visited the home of a Zulu village headman, we were stopped by his pack of Afri-Canis dogs. The dogs listened closely as the

man talked, obeying his every motion or word, keeping sharp eyes on their human master. The headman explained that he had another pack of dogs protecting his cattle herd as it grazed the highlands. He would see the cattle and dogs in a few days when they came to the lowlands for water.

We found Zululand to be a land of no fences and no firm borders, and soon encountered beautiful herds of Nguni cattle. We saw plenty of meat goats and cattle, but few sheep. Much of the land is rolling hills covered with thick grass, but not close enough to livestock grazing to keep it controlled, giving way to more arid territory, covered in thornbush. Men use packs of AfriCanis dogs to hunt wild pigs, hares,

AfriCanis dogs come in a variety of colors, but brindle dogs are common.

and jackals. We learned from our visits with the men that rabies is a significant concern for dog owners in Zululand.

Our African experience left me with a lingering perception matching that of other places we've visited that also have a long tradition tying livestock production and dog rearing. That perception includes a close relationship between humans and their animal partners, including some level of voice control of both livestock and dogs. The bond between the dogs and their humans, including a profound attentiveness and devotion of the dogs to their human handlers, was evident.

The AfriCanis dogs of Zululand are similar to the dogs that accompany nomadic people such as the Bedouin, Touareg, and other tribes in Africa—all are medium-sized sight hounds with a dual purpose. They have a close connection to their moving *morra*, protecting villages and encampments containing their people and associated livestock, while hunting for gazelle and hares with their human companions.

Africa's native dogs are built for coursing, with a behavioral bonus of a strong protective instinct. Africa's dogs defy the principle that a livestock guardian dog is a large animal without a strong prey drive—in the case of Africa's dogs, livestock guarding is a behavior rather than a physical form. Large guardian dogs like those that evolved in cold northern environments would seem little suited for Africa's hot climate.

A typical AfriCanis dog in Southern Africa.

Some question how a coursing dog—a sighthound, an animal with a strong prey drive—could be a successful livestock guardian. We needn't look any further than the culture in which the dogs are raised for the answer. Pastoral people around the globe have learned that if their pups are raised in the presence of livestock from a young age, the animals become protective. Their human shepherds select the dogs with the desired characteristics and correct behavior. Dogs that are too large or too small are not retained. Dogs that make few demands on limited resources are given preference. Dogs that kill or injure livestock, or bite people, in these pastoral cultures are not tolerated. Just as a Central Asian Ovcharka can initiate a vicious attack on a wolf approaching

its flock while behaving gently toward a newborn lamb, a sight hound can have a robust prey drive toward wild animals while maintaining strong protective behaviors toward its domestic livestock and people.

The advantage of using native dog breeds is gaining in popularity in Africa, with Cheetah Conservation Botswana discovering that local, mixed-breed dogs in Botswana (Tswana dogs) are effective livestock guardians, as well as cheap to acquire and maintain for low-income pastoralists. An added bonus is that the dogs are readily available, without requiring the intervention or involvement of organizations or government entities.

We found that stockgrowers we met in Africa have an intimate knowledge of the

AfriCanis dogs owned by Johan and Edith Gallant of South Africa.

animals they tend, and were deliberately selecting animals (both livestock and dogs) for individual matings for their most effective use and preferred behavioral traits. It's a lesson that has been reinforced wherever we've traveled where humans depend on their livestock and dogs for their livelihoods.

CHAPTER 16

PREDATORY PLOYS

A guardian chases a coyote away from the flock.

One fall morning as I sat at our kitchen table, papers scattered about its surface, I struggled to concentrate—balancing the checkbook, filing receipts, and reconciling accounts is a dreaded task for me. As I punched numbers, my subconscious mind heard the dog bark outside—an almost con-stant sound on the ranch, and nothing to be alarmed about. But eventually my ears perked up: that bark wasn't from any of our dogs.

The bark—sharp, higher-pitched, and insistent—was answered by the familiar booming voice of Rena, a female Akbash.

I threw open the back door to see one bold coyote brazenly challenging Rena to a chase, or to a brawl. Rena was young and cautious, and I could see her struggling to decide what to do. If she were to give chase, the lambs she would leave behind would be vulnerable. If she did nothing, well, the coyote was still there. When I stepped out the door to assess the situation, Rena made her decision. Leaving me with the lambs, Rena gave chase. I was mildly surprised when Rena did not disappear onto the folds of the Mesa, but returned quickly. About fifteen minutes later, I heard her bark again. A second coyote (with different marks and coloration) had appeared.

Rena seemed to have known that there were two coyotes trying to lure her away from the sheep. Perhaps they planned to ambush Rena when they got her on the Mesa alone. Or perhaps while one kept her occupied, the second could hit the lambs. Regardless, she didn't fall for their tricks. Careful glassing through the binoculars about an hour later revealed one of the coyotes lounging in the sunshine nearly a mile away, on a hillside overlooking our ranch. The coyotes kept their distance, but continued their watch.

The relationship between guardian dogs and various predators is complex, but we've noticed there is a great deal of watching

If not met with aggression, coyotes and wolves will stay near domestic livestock flocks, habituating guardians to their presence, and laying the framework for an easy future attack.

that takes place. Wolves are known to follow big game herds, with watching a part of their hunting sequence. It serves as a method of prey assessment, but I believe it serves a sort of habituation purpose as well. A predator that simply observes a herd from a nearby vantage point, on a regular basis, helps to habituate the herd to its presence (as researchers have noted in the French Alps[19]). When the prey species becomes habituated to the predator's presence, it enables easier subsequent attacks.

We've been able to watch the tactics employed by our local coyote population as the wily canines attempt to prey on our sheep flocks. Coyotes routinely test our guardian dogs and tease them. One winter day as I worked at home, I realized that Rena wasn't staying close to the house (our main flock was not located at our home place at the time, and was guarded by other dogs). Rena had no animals to protect, so I wondered what was keeping her busy. I found she had chased two coyotes out our back fence. The coyotes were attempting to lure Rena out of the pasture and onto the Mesa big game winter range, but I called her back. The coyotes could be trying to get Rena away from a safe place to attack the lone dog. Or they could simply be sizing her up, or habituating her to their presence. I spotted the two coyotes lounging around on the hillside on the Mesa later that afternoon, and fired a few shots at them to discourage their presence. They were too far away for me to get off a good shot, but at least I could make them know they were not welcome.

We moved the sheep flock back to our home, and with it came our other livestock guardian dogs. Every time our herd moves, it's an attraction for predators. It usually takes a few days for our dog team to dis-

place the coyotes from new range, but that year they had plenty of work to do.

One day, Rena and Luv's Girl arrived at the house at dawn, battle-weary and bloody after a night of conflict. The sheep flock was unscathed, and had been joined on their bedground by several hundred pronghorn antelope. Apparently the pronghorn realized that the safest place to be when there are predators on the prowl is with a guarded flock. Neither of the dogs was hurt badly, but Rena slept for almost nine straight hours in a spot just inside the door where she had collapsed upon entry. It was obvious from the frozen traces left on their neck manes that both dogs had been in physical conflicts with smaller animals that were trying to bite their throats. The smaller animals never succeeded, although Luv's Girl did have some swollen, bloody bites on her nose. Although coyotes are much smaller than the guardian dogs, these predators are very fast and agile. I was stunned one day to observe how much a skinned-out coyote carcass resembles the form of a greyhound dog—an animal that is built to run.

Because of the sheer persistence of our coyote threats, I had tried to keep one guardian dog kenneled at night—forced rest—while two others were on night duty with a small flock at the ranch. Rant, a Central Asian Ovcharka, had been doing a really good job while on duty, but he'd returned to the house nearly unable to walk a few times, suffering from exhaustion. The size of the coyote packs was dwindling, and I'm fairly confident that Rant was exerting lethal control. By the time lambing began, things were fairly quiet on our rangeland.

Lambing season in western Wyoming is my favorite time of year. I get up at 4 a.m. to drink coffee and eat breakfast before head-

A Central Asian Ovcharka keeping watch over ewes and newborn lambs.

ing outside for the morning check before the sun rises. Our lambing pasture in 2013 was a one-mile pasture around our house, and the sheep were scattered throughout. We'd had a cool and rainy week, so it was perfect conditions for lambing, with lots of green spring growth for the flock.

My checks always involve searching out each of our guardian animals. The dogs and burros manage to divvy up their presence in the flock as needed. The burros tend to work as a group, with all three burros surrounding ewes and newborns, grazing nearby but serving as a deterrent to disturbance.

This particular morning, the three burros were with three ewes and their four lambs, which were all born the day before.

This small nursery bunch remained surrounded by the calm presence of the three burros. The burros were formerly wild burros rounded up off the range in the desert southwest and then offered for adoption. The older pair is in their late twenties, and the young burro is about five years old. Wild burros have a natural aversion to members of the canine family, and will chase coyotes and foxes (or stray dogs), trying to stomp them to death. It took a period of time to get the burros and dogs to realize they are on the same team, but they now work well together.

In contrast to the burros, the dogs tend to work independently most of the time. If one bunch of sheep has a dog, the other dogs will seek out an unprotected bunch,

Burros and guardian dogs learn to work together to protect the herd from predators.

and all will patrol back and forth as needed. The dogs and the sheep have a very close relationship, but the bond is even stronger during lambing.

The sheep move several times during the day, and the dogs will move as needed, but if a ewe and lamb stay behind, so will one of the dogs. They have a keen perception of which members of the herd are most vulnerable. Sometimes during the day in the lambing season, the main flock will not have any dogs protecting them because the dogs remain with the animals needing more protection. The main flock moves together and does not include any ewes and lambs that can't keep up, so this is the group least in need of guarding during the day.

Our dogs seek out ewes as they begin labor, and one dog stays with any ewe as she gives birth, usually lounging on the ground nearby, but facing away from the ewe so it appears nonthreatening to her. If I come upon a ewe going into labor, or who has just had a lamb but is not yet in the company of a dog, I'll go find one of the dogs and haul it to the ewe, dropping the dog off nearby. The dogs understand the routine and happily comply.

One morning when I checked the sheep flock shortly after dawn, they were already on the move. I drove through them, noting that the herd had its burros, but no guardian dogs. Rena had already arrived at the house for breakfast, and Rant had been kenneled for a mandatory rest overnight,

Livestock Guardian Dogs

A guardian dog hides nearby as a ewe gives birth.

so where was Luv's Girl? I backtracked the flock and found Luv's Girl's bright-white body in the pale light, tail wagging as she stood to stretch. Nearby was a ewe with two newborn lambs. When the ewe had moved away from the herd bedding ground to go into labor, Luv's Girl had stayed with her, her presence nearby a comfort to the ewe, and a warning to predators. I dumped a dog food treat onto the ground for Luv's Girl and left the group on its own. Luv's Girl finally arrived at the house for food and water at mid-day.

I checked on the ewe and her twins a few times during the day, and just before dark I drove Rant over to them, placing him on guard duty for the night. He saw the lambs through the windshield, and eagerly exited the pickup truck. Rant took one look toward the new babes, then did his exaggerated eye roll, and began walking around the group, urinating on sagebrush. He was marking the territory around the new family, but never looking directly at them. When I returned to the house, I could hear his booming bark rolling across the hills. He was broadcasting his intent loudly across the landscape.

Usually within a day, these ewes and newborns seek out other ewes with newborns and form their own small nursery bunches. These small bunches stay together for a few days, sharing lamb-sitting duties before they naturally migrate back into the main flock.

One of the most important tasks the guardian dogs perform is to clean up afterbirth during lambing season, keeping our

Rant babysits a lamb that has become separated from the flock.

pastures clean of attractants that draw predators that might arrive for the birth matter, but soon seek out vulnerable newborns. Coyotes are, of course, our everyday threat, but avian predators are nearly as constant.

Our dogs abhor ravens, and have learned to harass them. We are fortunate not to have the golden eagle problems that plague many western sheep flocks, but we did once have a pair of bald eagles harass our flock and its newborn lambs. The dogs teamed up to harass the eagles enough to force them to move on. This might surprise those who don't consider bald eagles a threat to livestock. But the attraction of the red birth matter draws the keen-eyed eagle, whether it's a bald or golden, and it's not a huge leap to go from birth material to a newborn lamb. It happens with both golden eagles and bald eagles. The eagles retreated to perch on our back fenceline, but Rena took up fenceline patrol, harassing them until they gave up and retreated.

I receive so much pleasure in living with and working with the social creatures of my *morra* of burros, dogs, and sheep. I admire their toughness, their tenderness, and their ability to understand what needs to be done to take care of each other. It's our privilege to share life with such intelligent animals, and I continue to learn from them and trust their instincts.

One morning as I drove on my early morning check, I saw a ewe on her side with her head downhill in an unnatural position. At her head, feet resting on the inside of the ewe's neck, was Kit, a young livestock guardian dog. Anyone coming upon such a sight would logically assume the dog had taken the ewe down and this was a dire situation. But as I hurried forward, my presence disturbed the ewe, which kicked her back legs into action. The ewe jumped up, with a lamb that was in the process of departing her body then falling out to the ground. The dog jumped up as well, getting out of the way, tail wagging, happy the ewe had succeeded. What I had come upon was the dog showing concern for the laboring ewe, and the dog was elated when the new lamb made its appearance.

A few weeks later, I watched Kit come upon another ewe that was beginning labor. Kit sniffed the ewe's nose gently in greeting, and then calmly walked around to the ewe's backside, sniffing the area where the yet-to-be-born lamb's hooves were just appearing. Kit reclined on the ground just a few feet away, a quiet presence keeping the ewe company while she gave birth to two lambs.

Later that year, once lambing season was past, we moved the flock to new pastures about an hour away, and in the process, shuffling guardian dogs, acquiring a young male Akbash for the summer and fall grazing season, with Luv's Girl as his primary companion. Jim and I were headed out of the country for a few weeks, so the nearby ranch headquarters would be tending to our flock and taking care of our dogs, including our herding dogs. The time passed quickly, and the animals were fine in our absence.

It was a struggle to find suitable grazing for our sheep amid an ongoing drought, and after a month in one pasture, we needed to move the flock a few miles from their relatively safe pasture to another set of pastures in the foothills of the Wind River Mountains. As we exited the pasture, we found a freshly killed steer. A bear had ripped the hide and tissue off its face, and removed its lower jaw. Perhaps the pasture hadn't been as safe as we'd assumed.

It was just a short six-mile trail, but the threat level to the herd had increased to substantial: with coyotes the primary threat in the previous pasture, this area was home to wolves and bears. A pack of wolves killed a bull in one of the pastures two days prior. I was camping out with the herd, bedroll on a tarp on a green knoll, with the granite slopes of the Wind River Mountains spread out before me. My hope was that my presence, the buzz of my motorcycle, and my team of guardians, would work together to keep the flock safe. Of course, I was also packing plenty of firepower, but dreaded squeezing a trigger while trying to ascertain which is predator and which is prey or guardian animal, in a moment of chaos and terror.

Back at the house, our problem became a pack of five coyotes, although we usually only saw one or two together. This was a brazen bunch, with the adult female approaching near our house and outbuildings (but just out of my gun range), barking aggressively at Rena and our herding dogs. We'd been having coyote problems at our house since we returned after our absence. Since there were no livestock at home, and there hadn't been for several months, the coyotes had turned their attention to our dogs. Abe, our fourteen-year-old Bearded Collie, was a prime target.

The cause of the problem is fairly simple. We moved the sheep and their guardian dogs in early summer, so the dog niche was left essentially vacant. While we were absent, the herding dogs were gone also, and a local coyote pack that lives in the river bottom began to claim our place as home territory. Although the sheep flock

An Akbash approaches an aggressive coyote.

and its guardians remained on other range-lands, we returned home with two herding dogs and Rena.

We were notified of the coyote problem during our first night back, while we were sleeping in our own bed. The herding dogs were locked in the kennel behind the house. Since there were no sheep to be guarded, and I didn't want Rena roaming at night, she was also in with the small dogs. Sleeping with the window open, we were awakened by the sound of an aggressive coyote bark in our driveway, followed by the sounds of the dogs erupting, and Rena knocking out the top of the fence panel as she climbed over it on her way to challenge the coyote.

I threw on my robe and jumped in my truck, chasing Rena and the coyotes down the dark highway, and calling Rena back just as she was starting through the fence to the meadow on the other side. I hauled her back to the house and locked everyone inside so I could get some sleep. Jim fixed the fence so Rena couldn't get back over the top, but each night the coyotes would come back, trying to draw the dogs out. Abe had to sleep in the house, and most nights two-year-old Hud stayed inside as well.

The problem escalated enough that the coyotes were coming near our house in the early dawn almost daily. We were soon able to identify one of the main culprits as an adult female coyote. The bitch coyote would cross the highway from the west and make her way behind the house, barking aggressively and trying to call the dogs out. We wondered how many coyotes would be waiting to do an ambush, so Jim and I set up the game camera at a fencepost crossing a small draw, and photographed three coyotes in a single frame, so there were at least that many.

Rather than take the risk of shooting and missing, we let the bitch continue her aggressive behavior while we studied her habits. Soon she brought a second coyote with her, and both would stand just outside of shooting range, yowling and barking. The bitch seemed to know when Jim had a firearm, and would disappear into the sagebrush at those times. But he outsmarted her one day, hiding near the tractor and blowing on a predator call when the coyote approached to call out the dogs. The bold coyote came running in, never seeing him as he took one perfect fatal shot. We left her carcass near one of the trails along our back fenceline. It appears to have been enough of a disruption to the pack's social structure that the remainder of the pack became much less bold.

CHAPTER 17

HOME ON THE RANGE

A flock of sheep leaving the Wind River Mountains with its guardian dogs, shepherds, and herding dogs.

I sit perched atop a rock outcrop, listening to the quiet of the mountain morning. The flap, flap, flap of wings of an unseen bird. The snapping of deadfall as some large animal makes its way through the dark timber nearby. The barking chirps of a chipmunk, irritated to find visitors to its rocky habitat. The soft padding of Hud's footpads as he explores the nooks of the outcrop before reclining beside me. The occasion call of a rutting bull elk—a call answered by the whinny of a hunter's saddle horse.

We are waiting for the sheep. Today is the day that the first of the domestic sheep flocks will make its way out of the Bridger Wilderness of the Wind River Mountains, the start of their long migration south, reaching the high desert just north of the Colorado border the first week of December. Hud and I loaded up well before daylight for the sixty-mile drive so I can celebrate the flock's arrival.

Many countries of the world hold community events to commemorate the return of livestock herds from the high country in the fall, including a few celebrations in the American West, but my community is not one of those. Casting aside disturbing thoughts of the movement to rid public rangelands of seasonal grazing by livestock that would end this century-old tradition, I enjoy a picnic with my quiet herding dog as we wait for the sheep. A trout rises in the gentle waters of the small creek below us, and the loudest sound that we can hear is the crinkle of my sandwich bag.

Hud's mood today differs from mine. Initially enthusiastic, as cloud cover moves across the valley, drenching us in alternating sunlight and shadow, he hovers ever more closely to me, keeping his damp body pressed against my side as we watch the changing landscape spread out before us. We hadn't been back to this area since the year before at this time. On that rainy night while Hud and I slept next to our flock, we were visited by a pack of wolves. The livestock guardians kept the herd from harm that night, but Rena was nearly killed in the process. It's a night that will always be in our memory, and as I look at Hud, I wonder if he's having the same recollection. We're just a few miles from our camp site that night.

Or perhaps he's smelling and hearing things that I don't. We're in bear country,

with both black bears and grizzly bears present. There are wolves here, but we have no idea how many, or the size of the packs.

The silence of the morning is suddenly broken by the deep bark of a dog. Within seconds the joyful ring of a copper bell escapes through the trees. A rider on horseback appears, leading a laden pack string, accompanied closely by three herding dogs. I laugh in recognition when the first dog that comes into sight is one of Hud's brothers. Hud fusses next to me, wanting to race down off the rocks in greeting, but I tell him to stay put as I call out a "good morning" to the rider. The Nepalese herder replies in kind, and continues past, on his way to set up a tent and camp for the herders about to come off the mountain. The camp will be miles below, at the south end of these expansive meadows.

It would be another hour before I would hear the sound of an indecipherable roar in the distance. As the flock approaches, individual voices and bells can be heard, and I listened in fascination to the noisy mass of animals as the sheep moved through a narrow bottleneck of the trail, boxed in by tall pine trees on either side, and began to emerge into the open meadow before me.

The guardians were magnificent, moving like warriors with three in the lead, others scattered along the flanks, or moving inside the flock, barely indistinguishable among the moving sea of sheep. Trailing the flock was a shepherd on horseback, with his border collies pushing the flock forward, and several long-legged guardians trotting alongside.

Hud and I remained on the ledge above the herd, and as it moved directly below us, I realized that the guardian dogs hadn't yet seen us. I began to stand, saying "hello puppies," and the dogs reacted in surprise,

Livestock guardians moving with their migratory flock.

hackles immediately raised, and loud barks yipped as the dogs surged toward the rocks. I got to my feet and reminded Hud to stay right at my side. Four of the guardians slipped away from the flock and appeared at the top of the rocks within just a few seconds. I stood facing them, saying "Hey dogs," and "no" when one started forward, eyes on Hud. The dogs immediately backed off, acknowledging that a woman and her dog in the rocks posed no threat to the herd below. All but one of the dogs jumped back off the rocks and into the herd, but one adolescent male stayed behind. Hud

and I began to work our way off the rocks, trailing behind the passing flocks, and the young dog approached Hud, briefly touching noses before he turned to rejoin his herd.

It was a moment of joy and celebration, watching that united *morra* pass below us, knowing that these animals had shared months of living high in the rugged mountains, not just surviving, but thriving, while sharing the range with some of the largest carnivores on the continent. Each component of the wild ecosystem is present, with brave livestock guardians providing the link that unites it all.

A young livestock guardian dog investigates intruders near his flock.

A livestock guardian dog in Wyoming's mountains.

Home on the Range

ACKNOWLEDGMENTS

I thank all the shepherds, researchers, and dog advocates around the world who so freely shared their knowledge and experience of working guardian dogs; the Wyoming Animal Damage Management Board and the Wyoming Wool Growers Association for assistance with field research; colleagues who helped to guide our international journeys, including Yolanda Cortés and Juan Carlos Blanco (Spain), Atila and Sider Sedefchev (Bulgaria), Guvener Isik (Turkey), Matthew Berry (Lesotho), and Johan and Edith Gallant (South Africa).

As always, I am eternally grateful to my husband Jim for being my partner in all of life's adventures. Our souls have been blessed for the time we have shared with noble beasts.

APPENDIX: MISCELLANEOUS PRACTICES

A Spanish Mastiff with double back dewclaws.

Dewclaws

Some guardian dog owners have the back dewclaws on their dogs removed, especially the pendulous double dewclaws on hind legs often found on Great Pyrenees dogs, while others judge the quality of an animal by the presence of the back dewclaws.

A Central Asian Ovcharka pup with cropped ears and mid-length docked tail.

If back dewclaws are attached only by a flap of skin, the digits are not useable by the dog, and can hang up, tear, or injure the dog. In some cases, the dog owner may want to have these removed to prevent injury to the dog. Other dogs have back dewclaws (even double back dewclaws) that are firmly anchored to the bone of the leg. The dogs are able to use these digits for gripping.

Nearly all the livestock guardian dog owners we visited in Bulgaria, Spain, and Turkey view back dewclaws as important, and even have a preference for the double back dewclaws as a trait of purebred guardians.

Ear Cropping & Tail Docking

In their countries of origin, the ears on many guardian dogs are cropped, and the tails may be docked at the mid-way point. Some believe that cropping ears makes for better hearing, but the primary and most traditional reason for ear cropping is that cropped ears will not be grabbed and ripped during a physical altercation with a wolf or with other dogs.

In Turkey, most shepherds dock the ears of their dogs, but a few others do not. In Bulgaria, shepherds crop one ear of their Karakachans, purportedly based on the belief that the dogs will be able to hear better.

It has been reported that cropping and docking is usually done three to ten days after birth, before the puppies' eyes are open. In some countries, cropping and docking are performed during the first week of life; when the owner comes to see the litter and to choose which pups to keep; or when the

An Akbash female with intact ears (left), and a Central Asian Ovcharka male with cropped ears (right).

bitch brings the puppies to the camp, usually when they are around four weeks old.

The American Veterinary Medical Association opposes ear cropping and tail docking of dogs when done solely for cosmetic purposes. The AVMA encourages the elimination of ear cropping and tail docking from breed standards. AVMA adopted this revised policy in November 2008.

Ear cropping on guardian dogs has few similarities to ear cropping done on certain breeds like the Doberman Pinscher, in which the ear stubs are then taped into an upright position purely for cosmetic purposes. Many European countries have banned both tail and ear cropping.

It is recommended that breeders of guardian dogs in areas with large carnivores consult with their veterinarians about the benefits and consequences of ear cropping and tail docking their litters as they are born.

Neutering

Conservation organizations or others promoting the use of livestock guardian dogs often encourage spaying and neutering of the dogs so that the dogs are not distracted by the drive to reproduce. But that practice fails to recognize that livestock producers are in the animal reproduction business and may want to have sexually reproductive dogs in order to ensure the continuation of desirable guardian traits. A livestock producer may want to be able to breed his or her own guardian dogs either as replacement animals, or as added income. Livestock producers—as they do with the other domestic species they rear—learn to manage their breeding dog stock, kenneling or controlling females as needed in order to control breeding. Some pastoralists assert that castrated dogs can never reach their full potential, or become lazy after desexing.

The decision to spay or neuter a dog should involve a consideration of the increased health risks resulting from such procedures (such as joint disorders and cancers). Laura J. Sanborn, MS, wrote a comprehensive paper detailing the health risks that is easily accessed on the Internet, and is posted on the National Animal Interest Alliance website (www.naiaonline.org).[29]

Non-reproductive issues need consideration as well. Research suggests that spayed female dogs of some breeds tend to be more aggressive toward humans than intact females, while castration can reduce aggression in male dogs, which may hinder their effectiveness as guardians.

In our travels throughout Europe and Turkey, we talked to a few guardian dog owners who had neutered dogs because of roaming. All others kept their dog's reproductive system intact.

Feeding
Just as a pack of wolves will shadow a herd of elk, bison, or caribou as it moves across the landscape, a pack of livestock guard-

A mated pair of Karakachans wearing dangle sticks.

ian dogs does the same with its flock of sheep (or goats, horses, cattle, etc.). While the wolves seek out members of the herd to kill, the guardian dogs seek to protect the herd from any potential danger. Both the wolves and the dogs benefit from eating the manure left by the herd. This is natural, and may provide a biological advantage to the dogs, since the manure contains partially digested plant material, vitamin-rich bacteria, and other nutrients that can be lacking in other parts of its diet. In many undeveloped countries, guardian dogs, as well as other dogs, consume an assortment of feces, including those from humans.

Dog owners are often concerned about proper nutrition and feeding of their animals, and guardians must be fed well to ensure they are adequately fueled to do their jobs, but not overfed so they grow too fat, become lethargic, lazy, or unable to chase and travel with their herds. Breeders all over the world feed dog kibble where it is available, but this is a relatively new source of nutrition for many guardian dogs.

Most of the dog owners we met in Spain feed their dogs kibble, but supplement that with breads or meat, and even sheep pellets. In Tajikistan, the dogs are fed pita bread and scraps, while in Uzbekistan they are fed a porridge made by mixing whole wheat and coarse flour with water or milk.

The preparation of a porridge-type food for the dogs appears to occur throughout many areas of the world. The grain or flour mixture is always heated to aid in the dog's digestion. In South Africa, dogs are fed *mieliepap*, a porridge made from ground maize. We saw livestock guardian dog pups fed a heated grain-based porridge from a communal dish in Bulgaria, with milk or yogurt added as available. All the livestock owners we interviewed in Turkey and Bul-

garia fed *yal*, although all Bulgarian owners also fed the dogs raw meat as well. Many livestock producers in Turkey also fed their dogs wild boar meat.

Some dog owners in Turkey also bake biscuits or breads for the dogs. In areas where it is abundant, milk from livestock is fed to the dogs, as are eggs or leftover scraps and vegetables.

Wherever they occur, guardian dogs catch and eat small rodents, and eat carrion and afterbirth. In many regions of the world, the carcasses of dead livestock are cut open and fed to the dogs.

Dangle Sticks

In parts of Europe, laws are still enforced that require livestock guardian dogs to wear a dangle stick attached to its collar by a swivel to prevent the dogs from chasing game animals. The law has been in place for centuries and is advocated by hunters seeking protection for wild game animals.

Dangle sticks hinder the dog's movement, bumping into its legs when the dog tries to run. In addition to the potential of causing injuries to the dog's legs, the dangle stick hinders the ability of the dog to chase away predators. Some livestock guardian dog training manuals recommend using dangle sticks to discourage young dogs from playfully chasing livestock. The stick is used as a temporary training tool rather than a permanent attachment to the dog's collar.

Kenneling & Tethering

Some owners express concern about the possibility of kenneling guardians for a long period of time, such as during the winter months when sheep flocks have restricted movement. This is actually a fairly common occurrence in areas of Europe. Researchers Jeffrey Green and Roger Woodruff of the

U.S. Sheep Experiment Station in Idaho kenneled up to fifteen livestock guardian dogs at a time while the sheep were confined to a feedlot in the winter. They reported that "for most dogs, the bond to sheep remains and may even be intensified with periods of separation. The period of isolation in the kennel appears to enhance the dog's desire and enthusiasm for the freedom of being with sheep."[30]

Tethering is a management tool used throughout the world with working livestock guardian dogs. Sometimes the animals are tethered for a few days, a few weeks, or a few months. The number of guardian dogs used with a flock varies, so some dogs are kenneled or tethered when not guarding the flocks.

The reasons to tether dogs varies throughout the year, including:

- Young dogs that are not yet mature/trustworthy enough to work on lambing grounds;
- Male dogs that are fighting for bitches in heat;
- During fall sorting and shipping, when flocks are combined (to keep the dogs from being injured in conflicts, or run over by vehicles);

- When predator conflicts escalate, guardian dogs on duty are rotated, forcing days of rest for exhausted but dedicated guardians;
- When flocks are grazed near residential areas, the number of working guardians is sometimes reduced in order to minimize the risk of conflicts near inhabited areas;
- When a dog has been injured in a conflict with predators, tethering the dog at the shepherd's camp near the flock allows the dog to rest and heal under human supervision (the dog would be miserable away from the sheep—some guardians won't eat if they are kenneled away from their sheep);
- Females with new pups will sometimes attempt to move pups into dens in the wild, and the female must be tethered in order to keep the pups in a safe place.

Tethering or kenneling working guardian dogs can be an important management tool to help ensure the dog's safety and welfare—it does not equate to neglect. With migratory livestock herds, kennels aren't always an option.

Spiked Collar Construction

This collar is constructed of heavy webbing with a felt liner sewn onto the backing. This two-inch-wide collar is thirty-one inches long, with about six inches on each end that are free from the spikes. Each iron spike is riveted to the collar, and spaced two inches from the next spike, for a total of eight. The spikes are made of half-inch wide by an eighth-inch-thick iron strap, with sharp sheared points. The iron strap is seven inches long, and with two inches riveted flat to the collar, with the two two-and-a-half-inch spikes bent upward.

This twenty-two-inch-long, light link collar is made of a series of interconnected two-inch-long links. Each link is made of a six-inch-long piece of steel rod sharpened on both ends, and when doubled, the spiked tips are bent outwards. The image demonstrates how the final link is joined.

This is a heavier version of the light link collar just described. This twenty-eight-inch-long collar is made of interconnected links four inches long when connected. A three-inch strip of half-inch-thick wool felt is sewn on to provide a backing so the cold iron doesn't touch the dog's skin in winter. The felt is removed in summer.

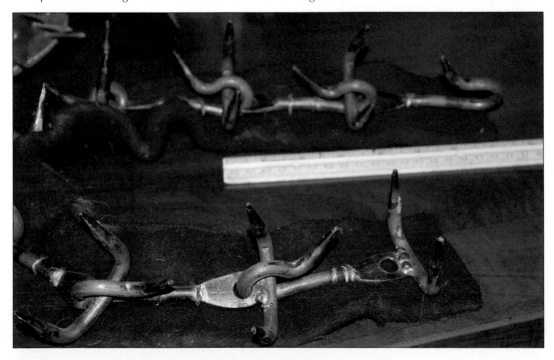

These one-and-a-half-inch-wide by one-sixteenth-inch iron strap collars are twenty inches long. The spikes are made of a half-inch by an eighth-inch-thick iron strap. The spikes bend outward at a length of two and three-quarters inches. The spikes can either be hot-riveted or welded onto the strap. Wool felting can be sewn on, and bells can be attached to help keep track of the dog.

These collars show the patterns of cuts made with sheet-metal shears. The felt has been riveted on to provide a soft backing. The chain link is welded on to allow the dog to be tethered, but not all of these collars have this feature.

These leather collars can be produced in any saddle or harness shop. Each is twenty-eight inches long, and two and three-quarters inches wide, with four to five inches free of spikes on each end. The collars were made to adjust it from twenty-two to twenty-six inches. This collar uses rows of one-and-a-half-inch broad-head roofing nails. No matter the length, the nails will bend over time and may need to be replaced.

A rangeland guardian in Wyoming wearing a spiked leather collar.

Livestock Guardian Dog Management Practices

Country	Producer	Species	Back Dewclaws	Crop Ears	Dock Tails•	Neuter	Feed••	Consume Carcass	Consume Afterbirth	Spiked Collars•••
Spain										
	Goya	Sheep	Yes/double	No	No	No	Kibble	Yes	No	No
	Rufino	Sheep	Yes/double	No	No	No	Kibble/bread	No	No	No
	Francisco	Sheep	Yes/double	No	No•	Yes	Sheep pellets/bread	Yes	No	No
	Carlos	Cattle	Yes/double	No	No	No	Kibble	No	Yes	Yes
	Juan	Cattle & Sheep	Yes/double	No	No•	No	Kibble/meat	No	No	Yes
	Paulino	Goats	Yes/double	No	No•	No	Kibble/bread	No	No	Yes
Bulgaria										
	Georgi	Sheep	No	No	No•	No	Meat/yal	No	Yes	Yes
	Daniela	Sheep & Goats	Yes/double	No	No	No	Meat/yal	Yes	Yes	Yes
	Miroslav	Sheep	Yes/double	No	No	No	Meat/Yal/Kibble	Yes	Yes	Yes
	Sider	Sheep & Goats	Yes/double	No	No	No	Meat/Yal	Yes	Yes	Yes
Turkey										

Livestock Guardian Dog Management Practices

Country	Producer	Species	Back Dewclaws	Crop Ears	Dock Tails•	Neuter	Feed••	Consume Carcass	Consume Afterbirth	Spiked Collars•••
	Kayis	Sheep	Yes	Yes	No	No	Yal/ Wild Boar	No	Yes	Yes
	Isparta	Goats	Yes	Yes	No	No	Yal	Yes	Yes	No
	Ali	Goats	Yes	Yes	No	No	Yal	Yes	Yes	No
	Kultu	Sheep	Yes	Yes	No•	No	Yal	Yes	Yes	Yes
	Memet	Cattle	Yes	Yes	No	No	Yal	Yes	Yes	No

• Some occur naturally in the guardian dog population.
•• Yal is a grain-based porridge cooked by shepherds.
••• Producers in Spain use spiked collars made of leather, while those in Bulgaria and Turkey use collars made from iron or webbing.

ENDNOTES

1   R. P. Coppinger, C. K. Smith, and L. Miller, "Observations on Why Mongrels May Make Effective Livestock Protecting Dogs," *Journal of Range Management* 38, no. 6 (November 1985): 560–61.

2   Jay R. Lorenz, "Diffusion of Eurasian Guarding Dogs into American Agriculture: An Alternative Method of Predator Control" (PhD diss., Oregon State University, 1990).

3   Hal L. Black and Jeffrey S. Green, "Navajo Use of Mixed-breed Dogs for Management of Predators," *Journal of Range Management* 38, no. 1 (January 1985), 11–15.

4   Ray Coppinger, Jay Lorenz, John Glendinning, and Peter Pinardi, "Attentiveness of Guarding Dogs for Reducing Predation on Domestic Sheep," *Journal of Range Management* 36, no. 3 (May 1983): 275–79.

5   Raymond Coppinger, Lorna Coppinger, Gail Langeloh, Lori Gettler, and Jay Lorenz, "A Decade of Use of Livestock Guarding Dogs," *Proceedings of the Thirteenth Vertebrate Pest Conference* (1988), Paper 43.

6   Jeffrey S. Green and Roger F. Woodruff, "Breed Comparisons and Characteristics of Use of Livestock Guarding Dogs," *Journal of Range Management* 41, no. 3 (May 1988), 249–51.

7   William F. Andelt, "Relative Effectiveness of Guarding-Dog Breeds to Deter Predation on Domestic Sheep in Colorado," *Wildlife Society Bulletin* 27, no. 3 (Autumn 1999), 706–14.

8   Inger Hansen and Morten Bakken, "Livestock-Guarding Dogs in Norway: Part I. Interactions," *Journal of Range Management* 52, no. 1 (January 1999): 2–6.

9   Alejandro González, Andrés Novaro, Martín Funes, Oscar Pailacura, María Jose Bolgeri, and Susan Walker, "Mixed-breed Guarding Dogs Reduce Conflict Between Goat Herders and Native Carnivores in Patagonia," *Human–Wildlife Interactions* 6, no. 2 (Fall 2012): 327–34.

10 Robin Rigg, "The Extent of Predation on Livestock by Large Carnivores in Slovakia and Mitigating Carnivore-Human Conflict Using Livestock Guarding Dogs" (MSc thesis, University of Aberdeen, 2004).

11 Linda van Bommel and Chris N. Johnson, "Where Do Livestock Guardian Dogs Go? Movement Patterns of Free-Ranging Maremma Sheepdogs," *PLoS One* 9, no. 10 (October 2014).

12 Brian Hare, "Dogs Use Humans as Tools," *Encyclopedia of Animal Behavior*, Beckoff, M. (ed.) (2004).

13 Michelle Morgan, "Mongolian Native dogs and the Cultural Heritage of Pastoral Nomadism," *Journal of the International Society for Preservation of Primitive Aboriginal Dogs* 20 (2009): 41–48.

14 Alikhon Latifi and Arunas Derus, "The Central Asian Ovcharka in Tajikistan," *Journal of the International Society for Preservation of Primitive Aboriginal Dogs* 21 (2009).

15 Zaur Bagiev, "Faithful and Fearless," *Journal of the International Society for Preservation of Primitive Aboriginal Dogs* 10 (2006).

16 Jeff Siddoway, letter to the editor, *Idaho Mountain Express*, June 27, 2012.

17 Ed Bangs, Mike Jimenez, Carter Niemeyer, Tom Meier, Val Asher, Joe Fontaine, Mark Collinge, Larry Handegard, Rod Krischke, Doug Smith and Curt Mack, "Livestock Guarding Dogs and Wolves in the Northern Rocky Mountains of the United States," *Carnivore Damage Prevention News* (January 2005): 32–39.

18 Robert Vartanyan, "Caucasian Mountain Dogs (Caucasian Ovcharkas) of the Northern and Central Caucasus Part 2," *Journal of the International Society for Preservation of Primitive Aboriginal Dogs* 26 (2012).

19 Jean-Marc Landry, Gérard Millischer, Jean-Luc Borelli, and Gus Lyon, "The CanOvis Project: Studying Internal and External Factors that May Influence Livestock Guardian Dogs' Efficiency Against Wolf Predation," *Carnivore Damage Prevention News* 10 (Spring 2014): 21–30.

20 Sider Sedefchev, "The Karakachan Dog—Continuation of an Old Bulgarian Tradition," *Carnivore Damage Prevention News* 9 (December 2005): 14–19.

21 Cat Urbigkit and Jim Urbigkit, "A Review: The Use of Livestock Protection Dogs in Association with Large Carnivores in the Rocky Mountains," *Sheep & Goat Research Journal* 25 (2010): 1–8.

22 Tatyana Mikhailovna Ivanova, "The Central Asian Ovcharka: On Some Problems of Preserving the Breed," *Journal of the International Society for Preservation of Primitive Aboriginal Dogs* 20 (2009): 3–11.

23 Carla Cruz, "Livestock Guarding Dogs from Portugal: A Review of Current Knowledge," *Journal of the International Society for Preservation of Primitive Aboriginal Dogs* 20 (2009): 11–34.

24 Carla Cruz, "The Estrela Mountain Dog—A Portuguese Icon," *The Shepherd* (July 2012): 26–28.

25 Ilker Ünlü, "Turkish Livestock Guardian Dogs," originally published on www.turkcobanko-pekleri.org (February 2008), accessed via www.kangaldogsinternational.org/pdf/TurkishLGDs.pdf.

26 Bryan Weber et al., "Movements of Domestic Sheep in the Presence of Livestock Guardian Dogs," *Sheep & Goat Research Journal* 30 (July 2015): 18–23.

27 Thomas Gehring, Kurt Vercauteren, Megan Provost, and Anna Cellar, "Utility of Livestock-Protection Dogs for Deterring Wildlife from Cattle Farms," *Wildlife Research* 37 (2010): 715–721.

28 Robin Rigg et al., "Mitigating Carnivore–Livestock Conflict in Europe: Lessons from Slovakia," *Oryx* 45, no. 2 (2011): 272–280.

29 Laura J. Sanborn, "Long-Term Health Risks and Benefits Associated with Spay/Neuter in Dogs," *Journal of the American Veterinary Medical Association* 224, no. 3 (Feb 2004): 380–387.

30 Jeffrey Green and Roger Woodruff, "The Use of Eurasian Dogs to Protect Sheep from Predators in North America: A Summary of Research at the US Sheep Experiment Station," *Proceedings of the First Eastern Wildlife Damage Control Conference* (September 1983): 119–124.